Making the Most
of the Anthropocene

Making the Most of the Anthropocene

Facing the Future

MARK DENNY

Johns Hopkins University Press

BALTIMORE

Johns Hopkins University Press
2715 North Charles Street
Baltimore, Maryland 21218-4363
www.press.jhu.edu

Library of Congress Cataloging-in-Publication Data

Names: Denny, Mark, 1953–, author.
Title: Making the most of the anthropocene : facing the
future / Mark Denny.
Description: Baltimore : Johns Hopkins University Press, [2017] |
Includes bibliographical references and index.
Identifiers: LCCN 2016049244| ISBN 9781421423005
(hardcover : alk. paper) | ISBN 9781421423012 (electronic) |
ISBN 1421423006 (hardcover : alk. paper) |
ISBN 1421423014 (electronic)
Subjects: LCSH: Human ecology—Philosophy. | Nature—Effect of
human beings on—Philosophy.
Classification: LCC GF21 .D46 2017 | DDC 304.2—dc23
LC record available at https://lccn.loc.gov/2016049244

A catalog record for this book is available from the British
Library.

*Special discounts are available for bulk purchases of this book. For
more information, please contact Special Sales at 410-516-6936 or
specialsales@press.jhu.edu.*

Johns Hopkins University Press uses environmentally friendly book
materials, including recycled text paper that is composed of at least
30 percent post-consumer waste, whenever possible.

Contents

Acknowledgments

First and foremost I must thank Vince Burke, longtime editor at Johns Hopkins University Press, for suggesting this book project and for giving me free rein in writing it. The Anthropocene notion has generated a lot of books, but Vince wanted one with shtick, and I wanted one that ranged wider than the norm, covering all the subject areas that impact upon our future. He got us what we both wanted, and here is the result; thanks, Vince—you know that Xark would have been proud of us. Also at Hopkins, I thank all the people who were involved in the acquisition, editing, production, and marketing processes, including Deborah Bors, Tiffany Gasbarrini, Catherine Goldstead, Hilary Jacqmin, Juliana McCarthy, and Gene Taft.

In the Old Country, Mike Lowe has for the last 40 years supplied a steady stream of astute, usually cynical, and always incisive observations about political and economic life, local and global, often glimpsed and transmitted through a beer glass. Closer to home, Ben Hoffman has provided me with Anthropocene contacts and references and with copious amounts of wine, along with the sage observations of a grizzled traveler around this wicked world. To both of these intelligent and worldly guys, I am truly grateful.

Making the Most
of the Anthropocene

Introduction

"Klaatu barada nikto." These are possibly the best-known words of science fiction (along with "Beam me up, Scotty" and "Open the pod bay doors, HAL"). They are from the 1951 classic film *The Day the Earth Stood Still*, in which the alien Klaatu, who looks suspiciously like Michael Rennie, urges the earthling heroine Helen to say these words to Gort, his all-powerful robot. The literal translation is, it seems, "Klaatu is dying but do not destroy Earth in retaliation." The movie is about the propensity of humanity to mess things up badly—the Cold War was under way and World War II was a recent memory at the time of filming—the premise being that, if humans are so hopeless, then destroying us would not be such a bad thing for the rest of the universe. In 2008 the film historian Steven Jay Rubin recalled an interview with scriptwriter Edmund North, who said that a better translation is "There's hope for Earth, if the scientists can be reached."*

More than 60 years after the release of *The Day the Earth Stood Still*, humans are still messing things up badly, though not just in politics. Today there are concerns about the economic and financial direction of the world, and where we are taking our climate. Our influence on the planet and its biomass has never been greater, and the vast majority of scientists now agree that much of the recent changes in climate are due to human activity. We cannot predict the details of what will happen to our climate in the coming decades and centuries, precisely *because* we have such an influence over it. Yet Klaatu is right: there is hope for us if the scientists can be reached. Today it is fashionable among nonscientists to denigrate science because it has been abused—nuclear weapons, environmental degradation—and yet if we are to

*This translation is given in the 2008 DVD *Decoding "Klaatu Barada Nikto": Science Fiction as Metaphor*. In 2008, a (ham-fisted, by comparison) remake of the same title appeared, in which computer graphics and concern for the environment replaced political metaphor.

make our future a tolerable one, we must overcome this prejudice and return to the sunnier view of science that prevailed in 1951.

We humans mess up things that are within our reach. The Earth's biosphere and climate are within our reach. Ergo we will mess up the biosphere and climate. This logic is simplistic, but unfortunately it is also compelling. The notion (that the planet's surface is sufficiently within our grasp that we are the dominant force in shaping it) now has a name: the Anthropocene age. For proponents of this notion, an alien—like Klaatu, more advanced than us, but unlike him, ignorant of our planet—could arrive on Earth and tell that its surface geology is being shaped by a dominant life form.

• • •

The idea of an Anthropocene age has been around for a few years and is a very important and serious one, with consequences. First, I will say something here about the importance and, second, about the seriousness. In much of the rest of the book I will address the consequences.

First, then, the importance of the subject has spawned many, many books and articles. Some of these are truly wonderful, well written and informative, while many others are—not. Good books include the excellent Schwägerl *cri de coeur* and the more personal book by Gaia Vince (see the bibliography). These authors express an almost desperate optimism about the future of humanity, which is overstated, in my opinion. They need balancing, and I have some other possibilities to offer—but sober with shtick, adding some sugar to take with the bitter pill. More than 50 years on from Paul Ehrlich's bestseller *The Population Bomb*, and longer still since Aldo Leopold's *A Sand County Almanac*, we live in a world little changed in terms of consumptive behavior. Perhaps this is because we are who we are, a species that loves the present more than the future.

Second, there is the seriousness of these issues. Many readers and experts doubt the necessity of the Anthropocene concept—that humanity has changed the physical world enough to merit the declaration of a new geological age—and indeed the word is a trendy catchall for a hotchpotch of environmental issues confronting the planet. I recognize the trendiness of the word (a book entitled *Anthropocene Bandwagons* would sell better than a book entitled *Bandwagons*), yet I don't think it should distract us from confronting the serious issues involved and the need to approach them from every angle. Much of the worthy literature to date is deadly serious, and consequently is rather heavy reading. Hence the shtick here. My approach to the subject has been called "quirky" by one reviewer. So be it: if starting this book with a 1951 science fiction movie is what it takes to stimulate thought and get a conversation going, then let's get down to it. You've read this far, haven't you?

There is plenty of science in the book that you hold in your hands, but it is presented in a palatable way—no math equations, no graphs or figures, purely a narrative text, brightly written with a serious message. I hope this science and my message—quirkiness notwithstanding—will be absorbed by a readership of intelligent nonspecialists interested in environmental matters and the world that their grandchildren will inherit. There are readers who will heed Housman's poetic advice from a century ago: "But take it: if the smack is sour, / The better for the embittered hour." For, like it or not, humanity has worse days ahead—survivable, but worse. I don't see it being avoided, but I do think some of us will be better prepared for it than others. Having a clue about what is going to happen next likely won't make much difference for me, because I don't have any kids. But if you do, and you would like a peek into their world, read on.

●　●　●

Much of what the Anthropocene age entails requires us to speculate about our future and, to quote the Bard, therein lies the rub. The Danish physicist and Nobel laureate Niels Bohr once said: "Prediction is very difficult, especially about the future." This lucid and amusing observation is from a man famous within the physics community for the opacity of his lectures and for his turgid prose, and it is therefore all the more striking. Examining the future is difficult to get right unless we are considering a simple physical system that evolves in time according to known physical laws. Show me a pendulum and I can tell you with confidence what it will do next. But complex physical systems do not exhibit such mathematical continuity, and hideously complex physical, chemical, biological, political, and economic systems such as we have made in our biosphere are impossible to predict with anything approaching certainty. Yet we must attempt predictions and call upon science, not Gort, to fix the road ahead.

• • •

The book is organized into 42 short chapters or essays. These chapters are mostly stand-alone, but they are heavily cross-referenced and are connected by occasional introductions, summaries, and pointers—a kind of internal Baedeker or guidebook. My glimpse of the Anthropocene age will develop intermittently in the early chapters, before being formed into a coherent picture later on.

Stratigraphy—the Top Layer

We need not dig very deeply into this very technical field; the top layer will provide us with enough material to know the lay of the land. Apologies in advance for the number of italicized technical terms—stratigraphy is like that. (In fact, my apology is not sincere; you need to see the technical nature of some of this stuff in order to appreciate the nature of the Anthropocene debate.)

Deep time is the long, long stretch since the world began. Geologists invented it ahead of physicists, and Darwin needed it for his theory of evolution to work. Deep time is so long on the scale of human experiences—a lifetime, the existence of a nation, or the duration of recorded history—that things seemingly fixed and permanent are seen to flow (rocks erode, continents drift, the sun brightens). The deep time of Earth's existence, or at least the time that has passed since its surface solidified, is divided by geologists into chunks. The chunks are then subdivided, just as time itself is measured out in years, days, and hours. The largest chunk of deep time is the *eon*. Each eon is divided into *eras*; each era into *periods*; each period into *epochs*; sometimes each epoch is subdivided into *ages* (also called *stages*).

For the last 541 million years the Earth has been in the *Phanerozoic* eon. The boundary between this and the previous eon (the *Proterozoic*) is set by the *Cambrian Explosion*, a short (in geological terms) extent of time in which the variety of life fanned out drastically and, in particular, hard-shelled creatures whose remains could fossilize came into existence. The Phanerozoic eon is divided into three eras, the latest of which is the *Cenozoic*, which began 65 million years ago with a mass extinction (see chapter 4)—most of the dinosaurs died out (though birds remained), to be replaced in many ecological niches by mammals. Cenozoic means "new life." The Cenozoic era is divided into three periods, like an ice hockey game; the latest is the Quaternary, which began 2.59 million years ago. The analogy with ice hockey is apposite because the Quaternary period began with the current ice age. The Quaternary period is divided into epochs, the most recent of which is the *Holocene*, which began 11,700 years ago with the last glacial maximum, considered to be the beginning of the most recent interglacial warm period. The ages into which the Holocene epoch is divided depend on which continent we refer to; they are given different names and arise at different times.

Note that the rationales provided here for stratigraphic boundaries fall into two types: biological and astronomical. Stratigraphers can tell the age of a stratum by the age of the fossils that are embedded within it (or the other way around—they can age fossils from the strata in which they are embedded). Astronomy enters the picture because the Earth's orbit around the sun is not quite a regular circle; it is an ellipse which precesses (the axis rotates) around the sun in the same direction as the Earth; also, the ellipse's eccentricity—how elongated it is—changes over time. These regular changes in orbital characteristics are known as *Milankovitch cycles*. Another astronomical factor is solar output: the sun's brightness changes. All these phenomena affect our climate and in particular, they are thought to be the main factor behind the ice ages. Hence the connection with stratigraphy.

● ● ●

So stratigraphers have carved up the Earth's time into eras and epochs and such like, based upon the physical changes to the surface that have arisen for astronomical reasons, or which can be identified by biological markers left in the rocks. The resulting temporal structure is called the *geological time scale*. The epoch we currently live in is the Holocene, as we have seen; this name was first proposed by an early geologist, Sir Charles Lyell, in 1833. The name was formally adopted by the International Geological Congress in 1885, and the surface of our planet has been passing through the Holocene epoch ever since.

Perhaps for not much longer. In May 2000 the Nobel prize–winning atmospheric chemist Paul Crutzen and ecologist Eugene Stoermer published an article suggesting that the surface and atmosphere of Earth had been so altered by humanity as to warrant a change to the geological time scale. They suggested that the Holocene epoch ended recently and that we have now entered the Anthropocene epoch. Rumblings along these lines had been made from time to time since the 1960s by various Earth scientists, but the article by Crutzen and Stoermer caught people's attention. That humanity had changed the face of the Earth and influenced the constituents of its atmosphere enough to cause climate change was becoming widely accepted—the question that geologists pondered, then and now, is this: are these changes detectable by stratigraphers, so that we can justify altering the geological time scale?* In chapter 3 we address this question and look at some of the dates proposed for the start of the new epoch.

•   •   •

There is some crossover between stratigraphy and climate change. That is, changes to the climate can become fixed in geological strata, but not necessarily or uniformly. To take an extreme example: if dinosaurs went extinct due to the effects of a large asteroid colliding with Earth, as many paleobiologists believe, then the effects of this collision were likely worldwide (dust thrown into the stratosphere reduced light levels across the globe, killing plants as well as dinosaurs). Strata around the world reflect this impact in, for example, the levels of the rare element iridium, which is more common in asteroids than on Earth. So an event such as asteroid impact affects both climate and stratigraphy. We have seen that there are other crossover phenomena: climate affects life on the surface, and some animals and plants fossilize and enter rock strata. Ice ages influence climate and can be detected in strata.

However, there are other phenomena that affect climate or stratigraphy but not both. Thus, geomagnetic flips—reversals of the Earth's

*See Crutzen (2002), Crutzen and Stoermer (2000), Monastersky (2015), Morelle (2015), Owen (2010), Steffen et al. (2011), and Stromberg (2013).

magnetic field that occur randomly (three times every million years on average)—are recorded in rocks that erupt as lava, through deep sea vents at tectonic plate boundaries, and then solidify. Linear patterns of magnetic material in rocks, aligned one way or the other depending on the direction of the geomagnetic field, provide a sequence that can be compared with those of other rocks to indicate relative age. This magnetostratigraphy has nothing to do with climate or climate change because it happens on the ocean floors. Experts can infer changes in the climate of ancient times by examining fossil pollen, but do such paleoclimatological clues affect rock strata? In fact, yes. Some of the geological ages are determined by fossil pollen distribution, but the effects vary from one continent to the next.

Humans affect climate, which may affect stratigraphy, or may not. Humans affect stratigraphy directly by mining activities, but is this disruption of the surface widespread enough to be significant to a stratigrapher? We need to look at the human influence on the climate and on the Earth's surface before we can decide. Officially there is not yet an Anthropocene epoch—but the concept has interested the International Union of Geological Sciences enough to set up a subcommission called the Anthropocene Working Group.* They will consider the evidence and then decide if we are now living in the Age of Man and if so, when it began.

 # Neptune versus Pluto

There has been a long-running fight between science and religion in philosophical circles of the western world from the time of the Renaissance. One by one, a number of different sciences have butted heads with the once-undisputed Christian theological version of how the universe in general, and our planet in particular, works. First up was physics; Round One in the seventeenth century took the form of a vicious debate over whether the sun orbited Earth or the other way

*See www.quaternary.stratigraphy.org/workinggroups/anthropocene, and the many references therein.

around.* Physics won; by the eighteenth century most people, including those within the established churches of Christendom, accepted that Earth and the other planets orbited around the sun. Then in Round Two it was the turn of geology: from its birth at the beginning of the nineteenth century, some geologists fought the established religious views on the age of our planet and the way that it formed and developed. Again, as we will see, the upstart science won the debate—today most people accept the geologists' views on planetary development. Round Three saw biology step into the ring, in the second half of the nineteenth century. The dispute concerned evolution—the development of life and in particular of humans. This debate overlapped with that of the geologists in that the age of the Earth proved to be a pivotal factor. The scientists, principally Darwin and his "bulldog" Huxley, eventually won the debate—most people today accept evolution in its modern, neo-Darwinian form, and not the biblical creation mythology.

The length of these disputes between science and religion points to the complexity of the issues under discussion and the intelligence and erudition of the proponents on both sides. The conflict took place at all levels, from the highest rational debate to the lowest violent brawling, and often involved tangential political issues that did not help to raise the tone. In the debate over evolution in Victorian Britain, for example, many of the opponents of Darwin were establishment figures who held the views of the state religion (the protestant Church of England), whereas Darwin held nonconformist religious views before losing his faith altogether. The evolutionists succeeded in winning over most Christians from their creationist position over the course of several decades because of the honest nature of the debate at the top level and because of the evidence they were able to bring to the table.†

---

*Vicious because it was unhealthy to disagree with religious orthodoxy at the time. Bruno was burned at the stake and Galileo was threatened with torture over this issue, among others. It is worth pointing out that both these men, and others such as Isaac Newton, who challenged orthodox beliefs of the day, held very unorthodox and (by modern standards) weird beliefs on all sorts of subjects, not just science.

†There are a few modern creationist holdouts, mostly in the United States. The level of their debate is, sadly, far below that of the original opponents of evolutionary theory, and seems to adopt the approach of lawyers rather than that of philoso-

Here I will say a little more about Round Two (i.e., about the struggle of geologists to establish their opinion on the formation of the Earth) because this matter is important if we are to have a better-than-superficial understanding of the ideas underpinning the Anthropocene age. Let me introduce to you the two champions of this contest. In the Neptunist corner we have Abraham Werner, a prominent geologist in the early 1800s, who was a professor of mineralogy at the Freiberg Mining Academy in what is now Germany. In the Plutonist corner we have James Hutton, an eccentric Scottish gentleman. The Neptunists believed that the world began as a mixture of minerals and water, that the minerals crystallized and formed sediments, and hence that modern rocks are all sedimentary. Plutonists saw vulcanism as the driving force of planetary development, and thought that sediments followed later after the igneous (volcanic) rock eroded and washed into the oceans. The names of the two groups (both sides had many champions and supporters) come from the Roman gods Neptune, ruler of the seas, and Pluto, ruler of the underworld.

An important distinction between these two paradigms of geological development is this: the Neptunists considered that the sedimentation process that formed rocks happened in the past and is now over and done with—there is no further geological development. The Plutonists believed that geological development is ongoing—it is happening today just as much as it was in the distant past, a view that was dignified with the longest word that you will find in this section: uniformitarianism.* Hutton wrote a book called *The Theory of the Earth* in which he outlined his beliefs, and in particular the necessity for his Plutonist argument of a very old Earth, much older than was generally believed in his day. Unfortunately, Hutton wrote badly, and so this book did not succeed in spreading his uniformitarian vision of the world. However, he had an articulate champion in John Playfair, a product of the Scottish Enlightenment who took the best bits from

---

phers: any tactic that gets the right verdict is justified, and evidence is useful only if it supports their view. For excellent biographies of Charles Darwin and his principal spokesman Thomas Henry Huxley, see Desmond and Moore (1991) and Desmond (1998).

*The longest words in the book are *paleoclimatological* and *magnetostratigraphy*, both with 19 letters. I thought you should know.

Hutton's book and argued for them intelligently and persuasively. Werner had the support of established churches because Neptunism was compatible with the Old Testament stories about The Flood. Also, the general belief in the early 1800s of a recent origin for our planet (in 4004 BC, according to a detailed reading of the Bible by a seventeenth-century Irish bishop, James Ussher) was not compatible with Hutton's view that the Earth was millions of years old.

James Hutton is considered the father of modern geology only in part because of his Plutonist views (which he developed rather than originated). Hutton read the rocks that poked above the surface near Edinburgh, where he lived, and his emphasis on strata and how they tell us about the long-ago developments of the planetary surface came to be accepted, slowly at first but then much more widely as evidence came in. Thus, to provide just one example, rock formations with volcanic intrusions through sedimentary layers are incompatible with the Neptunist viewpoint. Today stratigraphy—the branch of geology that studies rock layers and layering—is a sophisticated and mature discipline, and stratigraphers are the established authority that upstart proponents of Anthropocene ideas must convince if an Anthropocene age is to be declared. Of course the upstarts are not always right and the establishment is not always wrong, despite Hollywood conventions, and so we segue into the Anthropocene debate.*

•   •   •

Finally for this chapter, I would like to briefly opine about the triumph of science in the intellectual arena of the last four or five centuries. Science has to be rational because it seeks to describe Nature, which operates independently of human emotions, preferences, and foibles. That such emotional, biased, and foibled (to coin a word) creatures have created an intellectual edifice that trumps their own nature is quite remarkable. Scientists are human, and yet the way that science is properly conducted (through anonymous peer review of research, and funding that does not depend on monetary return) overcomes our self-

*For readable accounts of the origins of geology and the debate between Neptunists and Plutonists, see, e.g., Bressan (2010), Leddra (2010), and Press and Tanur (2001), chapter 4. For a charming account of Hutton's life and work, see Repcheck (2003).

analyzed weaknesses in the area of rational thought. The result is an understanding of nature that is becoming very deep indeed: there are two theories in physics that are very, very close to the truth, and for a scientist like me this fact makes them the most impressive constructs of human thought, ever—bar none. Both quantum electrodynamics (QED) and general relativity (GR) have been tested experimentally thousands of times, and so far they stand up to scrutiny very well. There is one QED prediction that is accurate to at least 11 decimal places.* What artistic, or theological, or philosophical, or environmentalist or biological description of nature can make quantifiable claims like that?

3 The Age of Man?

There are two questions here. First, has humanity changed the physical earth—its surface, atmosphere, and oceans—sufficiently to merit the declaration of a new geological epoch instigated by human activity? Second: if so, when did it begin? I will start with the second question, because it relates directly to the previous chapters and because it impacts on the more contentious first question.

• • •

The original start date for the onset of the hypothetical Anthropocene epoch, suggested by one of its proponents (Crutzen), was the late eighteenth century. This date marks the beginning of the first industrial revolution(see chapter 5), which has led to large and widespread physical changes to the Earth. In particular, from around the year 1800 coal-powered steam engines drove the industrialization of Great Britain and began a long process of carbon dioxide (CO_2) production, among other forms of industrial pollution, that imposed humankind's

---

*The prediction is of the electron anomalous magnetic moment. Intriguingly, the two theories—QED and GR—are incompatible on theoretical grounds, so at least one of them is wrong (i.e., not exactly true, albeit a close approximation) at the level of very small detail. In fact, to grind this metaphorical grist exceedingly fine, they probably both are.

carbon footprint on our planet. The level of CO_2 in the air has risen significantly and measurably since preindustrial times, as determined by analysis of air bubbles trapped in Arctic and Antarctic ice. Here we have a scientifically measurable marker for human influence on the *lithosphere* (the outermost layer of Earth's structure, the surface layer that is immediately below the atmosphere).

Another date that has been proposed by members of the Anthropocene Working Group is 1610. This precise time is again suggested by atmospheric CO_2 concentrations, here a noticeable dip in the level centered around that year. The dip is thought to have human origins, as follows. The discovery of the Americas by Europeans led to a great deal of exchange of goods, of animal and plant species, and of diseases, a process that has come to be known as the *Columbian Exchange*. Human diseases introduced from the Old World are estimated to have killed perhaps 50 million New World native people. Many of these people were farmers; following their deaths, agricultural land reverted to forest. Trees remove more carbon from the atmosphere than do crops, resulting in a dip in atmospheric CO_2 concentrations.

The rock strata so dear to stratigraphers will be marked in the future by a very thin but detectable smattering of radionuclides. They will be radioactive in a unique manner that reflects the fallout from nuclear weapons testing in the late 1950s and 1960s. This nuclear fallout peaked in 1964, and so some proponents of an Anthropocene age suggest that we take this date as the start of the epoch.* In terms of human alterations to the surface strata, a similar date—say mid-twentieth century—applies for the consequences of urbanization and mining. There are layers of concrete and brick, of disturbed and modified soils on farmland, and of polluted mudflats and estuaries. The postwar boom in population led to an increase in many parts of the world in human disruption of the surface—we have reshaped 77% of the land. Similarly, the construction of many dams around this time changed erosion rates, reducing the deposition of sediments at river mouths.

*See Prosh and McCracken (1985) for an amusing (in a disturbing kind of way) article about nuclear apocalypse providing the ultimate "golden spike" stratigraphic layer.

The global warming that has been an inevitable consequence of increased atmospheric CO_2 and hence of human industry (see chapter 14) has led to a rise in sea levels around the world. It will lead to greater sea-level rises in the future as the climate physics of global warming kicks in (the timescales are very long by human standards). Rapid sea-level rises are detectable geologically, as is another oceanic effect of increased CO_2 concentrations: acidification of the oceans, which causes coral reefs to dissolve.

Perhaps most significant in terms of permanent markers of human activity is the mass extinction of species (see chapter 4) and, importantly, the redistribution of plant and animal species across the land surfaces of the globe. Extinction rates today are 100 to 1,000 times the background (pre-human) rates, due directly to hunting or indirectly to habitat loss. Lost or redistributed species that leave fossil remains can provide a marker to future stratigraphers for a new epoch.

• • •

The Anthropocene Working Group has its work cut out because, while all of these suggested dates for the onset of the Anthropocene are plausible, none is the *golden spike* desired by stratigraphers. A golden spike is a unique, unmistakable, and widespread marker—say a uniform rock stratum of unique characteristics spread over the lithosphere. The influence of humans is growing as our population and its impact grow—some proponents of an early Anthropocene suggest a date several thousand years ago, when humans first developed agriculture, or mining. But the evidence is not widespread and varies significantly from region to region. In the future, as population increases (see chapter 11) and industrialization spreads, the marks made by our species on the lithosphere will increase. This increase will be steady, and so our presence on the planet will become more and more noticeable to an alien geologist from the future—so when will we be able to say that *this* date begins the Anthropocene? The problem for present-day earthly stratigraphers is that they deal with the past, not the future, and many of them are dubious about the whole notion of an Anthropocene epoch. Stratigraphers are the first among many to question the need for such an age—or, if it is inevitable due to human growth, they question setting the date soon. Let us wait a thousand years, they say, and then look back and declare that the Anthropocene began in

June 2192, when humanity indulged itself in a widespread nuclear war, or 2557, when all the land surface displayed strata of plastic,* or 2243, when the number of extinct fossilized species reached a minimum. They look at the long view: perhaps in a million years' time the current atmospheric carbon dioxide spike will not stand out from the noise.

"Is the Anthropocene an issue of stratigraphy or pop culture?" ask two geologists (Autin and Holbrook, 2012). The case for stratigraphy is the scientific one we have been making—that humans are leaving an unmistakable and measurable footprint on the lithosphere, and this footprint will probably become more widespread and deeper in the future. More cautious stratigraphers are dubious about environmentalist interest in defining an Anthropocene age, however. The environmentalists are not seeking scientific progress but rather want to foster broad social and cultural awareness of human-induced environmental changes. In this sense, they are seen as politicizing the debate, perhaps simply to draw attention to the impact of humans on the planet, and perhaps also with a longer-term goal of encouraging sustainable resource utilization. The Anthropocene debate certainly has kept environmental issues on magazine covers and in editorials of prominent publications (the *Economist*; the *New York Times*), though arguably global warming has never been off the front pages much for at least two decades.

In fact, a straw poll of stratigraphers has shown that over half think it is a good idea to look into the Anthropocene issue. One German stratigrapher, Stefan Wansa, presumably is not among them: "The proponents of the Anthropocene must confront the charge that they are not sufficiently familiar with the rules of stratigraphy" (Schwägerl and Bojanowski, 2011). Here is the main source of disquiet among stratigraphers, methinks. They are academics accustomed to rational debate of the abstruse technical points of a backwater subject who are

---

*In 1992, a shipment of 29,000 Chinese-made plastic ducks, frogs, turtles, and beavers that was bound for the United States fell overboard. The redistribution of these toys over time taught oceanographers about ocean currents; perhaps if this sort of accident becomes more common in future centuries, we will indeed have a stratum of plastic deposited around the world. See Adams (2011) for a report on the so-called Moby Ducks.

being thrown into the limelight by loud upstarts who are trying to hijack their subject for their own ends. Innocent stratigraphers are caught blinking in the headlights of publicity, while more media-savvy environmentalists are eating their lunch. Well, maybe. It is certainly the case that calling for an Anthropocene epoch has focused the attention of many people on the environment, on climate change, on industry and globalization, and on the future of the species that is marking the face of the Earth.*

The Anthropocene Working Group has decided, just before we went to press (2017), to recommend that we declare an Anthropocene epoch, which began around 1950. It was the radioactive fallout that did it, apparently, along with (I kid you not) the ubiquity of chicken bones strewn around the globe. Several years will elapse before this recommendation becomes official and the geological time scale is altered accordingly.†

 4 Martha

The dodo was a super-sized flightless pigeon endemic to the island of Mauritius, off the east coast of Madagascar in the Indian Ocean. It went extinct sometime in the latter half of the seventeenth century, probably as a result of human activity (a victim of hunting or of predatory species introduced to Mauritius by European sailors) and is today the icon of extinction in many parts of the world. "As dead as a dodo," we say. In North America another pigeon is iconic of extinction, and more poignantly so because we know much more about its demise.

Passenger pigeons were migratory birds (their name comes from the French *passager* for "passing" or "voyager") that formed huge flocks on their migratory journeys and at nesting sites. It has been estimated that

---

*The vast and varied literature on the stratigraphic case for an Anthropocene epoch includes the following useful articles: Autin and Holbrook (2012), Corneliussen (2015), Kolbert (2010), Monastersky (2015), Nijhuis (2015), Schwägerl and Bojanowski (2011), and Zalasiewicz et al. (2010).

†The recommendation of the Anthropocene Working group has been reported widely; see Amos (2016), Carrington (2016), St. Fleur (2016), and Mooney (2016).

they were among the most numerous of birds in the world before the nineteenth century and constituted between 25% and 40% of the total bird population of the United States, numbering perhaps 3 to 5 billion in total. Alexander Wilson, a prominent American ornithologist of the early 1800s, estimated one flock to contain over two billion birds. That is about 10 times the total number of wild pigeons (rock doves) in the world today. Flocks were so big that they frightened people; they would be a mile wide and take hours to pass overhead, and were very noisy. Audubon wrote about a flock that he encountered in 1813: "The air was literally filled with Pigeons; the light of noon-day was obscured as by an eclipse; the dung fell in spots, not unlike melting flakes of snow; and the continued buzz of wings had a tendency to lull my senses to repose."* There were many such accounts of these massive flocks. To say that passenger pigeons were "numerous" in the mid-1800s would be an extreme understatement, like referring to Attila the Hun as "confrontational."

Passenger pigeons went extinct at 1 p.m. on September 1, 1914.

They were good eating. Passenger pigeons were hunted to extinction, aided by the new technology of the second industrial revolution— guns, trains, the telegraph (see chapter 5)—and were sped on their way by a fragile ecology that made them vulnerable. There were no laws restricting hunting these birds; they were easily netted, but mostly shot. They were killed for private consumption or for sale. Hunters would be telegraphed the location of a large flock and would kill millions, stuff the birds into barrels and send them by trains to cities where they were sold in open-air markets or to fine restaurants. As a result of the slaughter, passenger pigeon numbers leveled off in mid-century and then went into a steep decline from 1870. They were almost unknown in the wild by 1890.

Too late, a few people—and then many across the United States— saw that extinction was a real possibility. A few survivors (passenger pigeons, not people) were captured and kept in zoos. Martha (named after the first First Lady), the very last passenger pigeon, was kept at the Cincinnati Zoological Garden with a few others. The last two males died there in 1910. Caretakers offered $1,000 to anyone who could supply a mate for Martha, but this offer was never taken up. Over

*Audubon, J. "On the Passenger Pigeon," in *The Birds of America*, 1827.

the period 1909–1912 the American Ornithologists Union offered $1,500 to anybody who could locate a nest in the wild—but it was too late. Martha suffered from an apoplectic stroke which she survived, several years before her death, but from which she never fully recovered. She had difficulty reaching her perch, so caretakers lowered it for her. Zoo visitors were disappointed that she did nothing but sit on her perch, hardly moving. She was twenty-nine when she died. What a contrast between her treatment and that of her predecessors.

Martha's body was immediately stored in ice and transferred to the Smithsonian Institute where it was skinned and preserved, and where it remains to this day. Her death raised public interest in creating strong conservation laws; the centennial of her death was marked by many head-shaking articles in the media. One team of scientists is trying to *de-extinct* the passenger pigeon by genetically engineering related species of birds to exhibit passenger pigeon traits. The closest extant relative is the mourning dove, which resembles the passenger pigeon in shape and coloring.*

• • •

According to paleontologists, the history of life on Earth has been punctured by five episodic *great extinctions*; these are brief periods during which a significant fraction of known species—at least half— are driven to extinction due to some natural catastrophe. These five events (which may have taken millions of years each to unfold, as several clusters of extinction events rather than a single cataclysm) are named from the geological time scale boundaries that they define:

Ordovician-Silurian. Around 445 million years ago, this is the extinction that saw off the trilobites—most of its victims were sea creatures. It was likely brought on by a climatic cold snap.
Late Devonian. About 270 million years ago. Three-quarters of species died, mostly in shallow seas.

---

*I relied mostly on the following centennial books and articles for this section on the passenger pigeon: Avery (2014), Harvey and Newbern (2014), Rosen (2014), and Zimmer (2014). See also the Smithsonian website www.si.edu/encyclopedia_si /nmnh/passpig.htm.

Permian. 252 million years ago. Fully 96% of marine species and 70% of terrestrial vertebrates died off, probably from massive volcanic eruptions that spewed vast amounts of carbon dioxide into the atmosphere. This mass extinction is the only one to include many insect species.

Triassic-Jurassic. 201 million years ago. Brought on by climate change, or perhaps volcanic activity—take your pick.

Cretaceous-Tertiary. 66 million years ago. Also known as K-T or Cretaceous-Paleogene, this is the extinction that dispatched the dinosaurs (and also ammonites). Caused most likely by a large asteroid (there were also the humongous Deccan volcanoes at this time, which helped).

It is difficult to be confident about the percentages of creatures that went extinct, due to the imperfect nature of fossil remains. First, only creatures that got stoned—fossilized—can be counted, because the others have left no record. Second, sedimentation that leads to fossilization tends to occur under the sea and in coastal regions, so the fossil record is biased in favor of marine creatures; we know little about mountain dwellers from earlier ages, for example.

Here is why these significant events in the history of our biosphere get into a book about the Anthropocene: there is likely a sixth great extinction under way right now, probably caused by human activity. You may be forgiven for not having noticed, if you do not read very widely, because of the distinct look of bloated corpses littering our fields and oceans. However, extinctions are not really like that. They occur rapidly on a geological timescale, which is to say, very slowly on a human scale. Even though the rate of extinctions today is at least 100 times the background rate (another number to be a little wary of, because it also relies on the fossil record and because it is difficult to establish the background rate), we barely notice it.*

---

*The Discovery Channel series *Lost Animals of the 20th Century* includes the following partial list of victims of the sixth great extinction from among the animal kingdom, to add to Martha and her clan: Arabian ostrich, Arizona jaguar, Bali tiger, Barbados raccoon, Barbary lion, bulldog rat, California grizzly, Cape red hartebeest, Cape Verde skink, Caribbean monk seal, Caspian tiger, Culebra Island parrot, Dawson's caribou, Falkland Island fox, Florida black wolf, gastric brooding frog, Greenland reindeer, Grévy's zebra, Guam flying fox, heath hen, ivory-billed

Humans are to blame partly because we have caused, inadvertently on the whole, an accelerated dispersal of species beyond their natural range, and these invasive species have killed off indigenous animals and plants—an example is the brown tree snake, which, when introduced to Guam by humans, promptly set about wiping out most of the native bird species. Another aspect—perhaps the main one—that points to a human cause for the current extinctions is habitat loss, and a third is climate warming. A 2007 IPCC (Intergovernmental Panel on Climate Change) report stated that between 20% and 30% of plant and animal species are at increased risk of becoming extinct this century, if the planet continues to warm at its present rate and creates a 2°C increase in average temperature over the preindustrial level. Again, the numbers are disputable, and have been disputed in both directions, greater and lesser.

The sixth great extinction announcement in the scientific literature led, predictably and understandably enough, to a slew (an unfortunate pun—apologies) of newspaper and magazine headlines along the lines of *We're Doomed*. Lurid speculation about the world that humankind and the remaining fauna and flora will be living in, say in the next century—and which we will have brought about as a result of our environmental folly—is natural but is difficult to take at face value. In much of the rest of this book I will set out where I think we are heading and why, but it will not be to extinction. The scientific claims that we are in the midst of a mass extinction event ("event" on the geological time scale) are quite convincing, and the follow-on claim that we are the cause of it is plausible. In a sense, it would be good news if a human cause is confirmed, because then there is the possibility that we can do something about it, but here I am less optimistic than some authors, for reasons that will later become clear. Humankind is directly or indirectly killing off many species today. Our descendants will not be able to benefit from these species, either as medicinal sources (Edward Wilson is fond of pointing out the potential pharmaceutical value

---

woodpecker, Japanese wolf, June sucker, Kamchatkan bear, Kona giant looper moth, Labrador duck, Mexican silver grizzly, Newfoundland white wolf, New Zealand grayling, Palestinian painted frog, paradise parrot, pink-headed duck, Portuguese ibex, red owl, Round Island boa, rufus gazelle, Saint Helena giant earwig, Schomburgk's deer, Syrian onager, Tasmanian wolf, Wake Island rail.

of many rare or recently discovered plant species) or as creatures to be looked at and appreciated for their beauty, just as we have been denied the chance to see giant flocks of passenger pigeons transform the sky.*

• • •

So we have changed the face of the planet, though perhaps not enough to satisfy stratigraphers—let's wait a millennium or so and then they can put a finger on a calendar date. There will very likely be an Anthropocene epoch, even if we do not yet know when it began or will begin. Humans are already adversely influencing biodiversity, mainly through habitat loss. We may be able to green up a habitat that we have degraded, but losing a species is permanent.

An industrial interlude now, to show where the modern world—its economy, technology, climate, and environment—came from, and then back to the Age of Man—how many of us will there be, during the Anthropocene?

⑤ Industrial Revelations

In *The Day the Earth Stood Still*, the Armageddon robot Gort stopped all the machines of the world from moving—hence the movie title. So it seems that Gort's capabilities included the ability to sense technology. Let us imagine a more benign scenario than that played out in the movie. Perhaps earlier in his career Klaatu had been an anthropologist, and his robot's function was to sense the technological development of a species through time. Let us imagine that Klaatu and Gort visited Earth to measure our industrial revolutions—periods of rapid engineering development coupled with the wide deployment of the resulting new technology within society. In particular, Klaatu considered that an industrial revolution consists of more than just a period of gradually rising technical sophistication during which machines

---

*For popular and technical accounts of great extinctions, particularly the sixth, see Ceballos, Ehrlich, and Ehrlich (2015), Ceballos et al. (2015), Drake (2015), Kolbert (2014), MacLeod (2015), Plumer (2014), and Zalasiewicz (2015). There is now evidence that the number of animals (as well as the number of animal species) is falling fast: down 58% between 1970 and 2012. See Doyle (2016).

evolve to perform better. To be a revolution, a period of development must lead to new machines that can perform tasks previous machines could not do at all, as well as to new machines that perform existing tasks better.

Gort shows Professor Klaatu that Earth's first industrial revolution (henceforth IR1—let me use the notation IRn for the nth industrial revolution) occurred in a group of islands off the northwest coast of Europe, from around 1780 AD to 1850 AD. This was something of a surprise, because Great Britain seemed at first glance to be an unremarkable place by other measures of civilization. IR2 consisted of the spread of the new technologies to other countries such as Germany and the United States, in the second half of the nineteenth century and the beginning of the twentieth. IR2 was more than simply a dissemination of IR1, however—important new industries were developed. IR3 occurred later, and it began around 1990 in the United States. This digital revolution is continuing today.

Here is a summary of the revolutionary hat trick of human industry, which has changed the quality and quantity of life on earth for many (but not yet all) of its inhabitants, and has changed the face of the planet.

●   ●   ●

But first, it is reasonable to ask if such revolutions are necessary, in order to increase the prosperity of a nation and its people. The answer from both experts and the evidence of history seems to be "yes." The first words of Phyllis Deane's venerable and excellent history of IR1 are these: "It is now almost an axiom of the theory of economic development that the route to affluence lies by way of an industrial revolution." What did the first such revolution consist of, and why did it start in Great Britain during the last quarter of the eighteenth century?*

*The quote is from Deane (1965). The names by which Britain is called are many and varied and lead to considerable confusion, even among the British and their Irish neighbors. The geographical name for the island group is *British Isles*; the biggest of these is *Great Britain,* and second biggest is *Ireland.* Politically, the British Isles are divided into two countries, the *United Kingdom of Great Britain and Northern Ireland* (UK for short) and the *Republic of Ireland.* It gets more complicated when we consider the states within each country, but we won't go there. Thus I was not quite correct when referring to Great Britain as the source of IR1,

There were many factors coming together in Britain at that time, all of which were necessary precursors for what would become IR1. This revolution was not a single planned event, like a political revolution—it happened spontaneously and without the benefit or hindrance of any direct government action, and once it got started it became faster and bigger, like a chain reaction. The precursors were:

- financial infrastructure
- laissez-faire economic outlook
- a culture of innovation
- transport infrastructure
- access to raw materials
- cheap labor

The financial infrastructure consisted of a well-developed system of banking and the availability of seed capital to fund new companies and industries—I suppose we would call them "blue-sky start-ups" today. The laissez-faire economics were newly fashionable at the time. A raft of laws repealing old practices that stifled development and free trade had been passed (for social reasons, not to stimulate a revolution) and the Scottish economist Adam Smith had just published his magnum opus *The Wealth of Nations* in 1776. The culture of innovation in some ways followed from the first two factors; British scientific and techno-logical development had from the seventeenth century followed differ-ent paths from those in the rest of Europe, and made better progress as a result. For example, scientific progress on the continent was largely at the whim of dilettante monarchs, who set up astronomical observa-tories and funded research into the fundamental sciences and mathe-matics, for personal interest or for prestige. In Britain, however, there was a lot of free money (that is, free of royal patronage) provided by private individuals who wanted to pursue research or who wished to become rich by improving some aspect of trade or manufacture. At the theoretical end of the scale there was the Lunar Society, a body of intelligent men (many of them political or religious nonconformists)

---

because part of it took place in Belfast, which is in Ireland. On the other hand, it would also have been incorrect to refer to the UK because, at the time of IR1, it was a different country: the *United Kingdom of Great Britain and Ireland*. Confused?

who met every full moon to discuss scientific matters: Erasmus Darwin (grandfather of Charles), Josiah Wedgwood, Matthew Boulton, James Watt, Joseph Priestley and others—some eminent scientists themselves, others inventors or doctors or entrepreneurs. At the business end of the scale, literally, were the brash, confident, ruthless, and capable men who either kick-started IR1 or who emerged from it and pushed it further and faster—the same Matthew Boulton (developer of Watt's steam engines), Abraham Darby, and later Henry Cort and John Wilkinson (iron masters), Henry Maudslay (machine tool manufacturer), and Isambard Kingdom Brunel (builder of innovative bridges and ships—indeed, of anything big), to name just a few.

The transport infrastructure in Britain at the end of the eighteenth century consisted of newly built roads and canals, newly invented railroads, and a large merchant marine, along with a navy to protect it. Sea transport was always important for an island "nation of shopkeepers," as Napoleon is said to have called the British.* Clearly, reliable and affordable bulk transport is necessary to move goods to market, to exchange ideas, and to obtain raw materials. The raw materials that Britain needed to power her new industries were water (at first the main source of power was waterwheels) and then coal (for steam engines); both of these were readily available. As was cheap labor: IR1 is infamous for the social dislocation and hardship that it caused, as poor country peasants flocked to urban slums to become the poor working class. These millions lived amid pollution and squalor, and depending on your political inclinations were either pulverized by their factory-owner oppressors (according to that product of the industrial revolution, Karl Marx) or were the victims of heartless employers in an uncaring society (according to another famous product of those times, Charles Dickens). Infamous this hardship and squalor was, and very real, yet living standards for the workers rose steadily.

It is important to convey the bootstrapping, nonlinear nature of IR1. For example, steam engines were initially used to pump water

*In fact, it was probably another French revolutionary (de Vieuzac) who used this term derogatorily to describe his British enemy. A very readable account of the Lunar Society is that of Uglow (2002). For more about different aspects of IR1, see, e.g., Denny (2007b) chapters 4 and 5, Denny (2013) chapter 2, Fara (2002), Mason (1962), Raymond (1984), Usher (1988), and Weightman (2010).

out of coal mines, thus increasing coal production. The coal was used to make iron for constructing (among many other things) steam engines. These engines were used to haul coal out of mines, thus increasing production, and transport it around the country, thus increasing sales and powering other industries. Another example: improved transport meant reduced stock-holding costs, reduced wastage of perishable products, and reduced prices at the point of sale. All of these improvements led to economic growth and so greater use of transport, which then expanded to increase carrying capacity. By the middle of the nineteenth century, Britain was making most of the world's iron and steel, producing most of the world's cotton clothing, and making and exporting most of the world's mechanical power in the form of steam engines for industry, ships, and the rapidly expanding railroads. Matthew Boulton said on many occasions, as he guided visitors around his steam engine factory in Birmingham: "I sell here, Sir, what all the world desires to have—Power" (Uglow, xi).

● ● ●

Much of the world soon got it. IR2 saw the spiraling growth of innovations and manufacturing that characterized Britain in the first half of the nineteenth century spread to Belgium, France, Germany, Japan, the Netherlands, and the United States, in the second half of the century (and persist until the outbreak of World War I). Economic historians reckon that there was something of a lull in the hectic pace of economic growth sometime in the 1830s and 1840s—more accurately a pause or slackening of pace—that justifies separating the two revolutions. In fact, even without such a hiatus, they would have drawn a distinction, because the two phenomena showed significant differences. IR2 changed not only the rate of production but also the organization of production: factories as we know them, with production lines powered by electricity, were invented. Mass production led to cheap goods and so the benefits of the growing economies fed down to the ordinary working people much more than was the case earlier. Purchasing power of wages increased, and consequently living standards rose.

Mass production lowers the cost of each unit produced—this is why ordinary people benefited. Mokyr (1998) provides the analogy of storage boxes. Suppose a factory makes cubic boxes that are 1 foot on each side for $10. By doubling the size, the storage volume increases by a factor of 8, whereas the amount of material needed to construct the box (and the production cost) increases by only a factor of 4. Therefore, cost per unit volume of storage halves. Hence, increasing the production of storage volume reduces unit cost. Now combine increased production with new methods of manufacture and you have a revolution—thus, for example, Henry Bessamer's process for making steel coupled with the newly-invented blast furnace is reckoned to have kick-started IR2, leading to cheap, mass-produced steel for the first time in history.

Mass production required large factories and corporations, and many of the world's big companies arose during the period of IR2. In the United States, these included Carnegie Steel, Dupont, Ford Motors, and General Electric. Britain kept pace with the two principal burgeoning nations at the beginning of this period, around the mid-1800s, but then handed over the baton to the United States and Germany. America had more people and land than Britain and an unfettered attitude to business that saw, for example, more miles of railroad track laid in the 1880s than in any other country or any other decade. IR2 created the worldwide growth of railroads. It also engendered the establishment of the chemical industry, and here it was Germany that took over the lead. The older British way of progressing was via talented individuals, often without much formal education, making empirical progress to understand a subject—say inventing a new artificial dye—and then capitalizing on it. The new German method was to learn by more gradual but surer steps made by men with formal training. Much of the modern field of organic chemistry was first understood by Germans at this time—men such as Liebeg, Wöhler, Bunsen, Gmelin, and von Hoffman. Theirs are hardly household names to you or me, but they are famous to any chemist today, because they established a field that very quickly led to the markets of the world being filled with new chemical products: the first manmade aniline dyes, new explosives such as dynamite, fertilizers, the first plastic (celluloid), and pharmaceuticals.

Steel, railroads, production lines, chemicals. There are many other processes and products that were first understood and then harnessed for human use during the period of IR2. The most important is electricity—electrical power reached factories and households, with obvious ramifications for the quality and quantity of life for people in industrialized countries (electric motors, electric lighting, etc). Power is moved from one place to another much more easily and quickly in electrical form than in any other form (water, steam, coal, oil), and the first modern power stations were built during IR2, from which electrical power was distributed far and wide. At the center of these power stations were steam turbines, developed during this period.

I will discuss just one more of the many other inventions and products that emerged during the second half of the nineteenth century,* and just one of its consequences—wholly unforeseen at the time. Refrigeration was developed in a German brewery in the mid-1800s to aid brewers in distributing their perishable product more widely. Refrigerated railroad cars and then household fridges followed. The distribution of food was made much easier and the wastage reduced. Beyond these obvious changes were more subtle but far-reaching consequences. Because beer could be spread farther and wider than before refrigeration, competition between breweries increased. To survive, brewers had to combine with or take over other breweries. Thus the number of breweries across the world dropped sharply from the end of the nineteenth century—a process that continued until the 1970s—until most of the world's beer was made by only a couple of dozen huge macrobreweries by the middle of the twentieth century. Here was a lesson for the future: unintended consequences of benign inventions.†

● ● ●

Environmental degradation is one obvious consequence of the first two industrial revolutions. See chapter 8 for a peek at the effects of early industrial pollution. This pollution arose from the sources of energy that were used in factories as well as from the manufacturing processes and products of those factories. The energy that was har-

---

*Mass-produced bicycles, vulcanized rubber, communications cables, the telephone and telegraph, paraffin and naphtha, petroleum and automobiles, internal combustion engines, lubricants, mass-produced paper, smokeless powder, mass-produced rifles, skyscrapers. For a nontechnical summary of the electrification of society, see Denny (2013) chapter 3; the effects of refrigeration on the brewing industry is made clear in Denny (2009) chapter 1.

†For more on IR2, see, e.g., Landes (1969), Misa (1995), Mokyr (1998), Nye (1992), and Smil (2005). Another unintended consequence of mass refrigeration was depletion of the ozone layer, as we will see.

nessed to first get IR1 up and running was gravitational: waterwheels were turned by flowing water. By the end of IR1, the energy was chemical, and came from coal: the hundreds, then thousands, then tens of thousands of steam engines that powered locomotives and looms were fired up by burning coal. Coal-powered steam engines still provided the drive at the beginning of IR2, but at the end the energy came from oil: when processed into gasoline and diesel, the energy was converted into mechanical power via millions of internal combustion engines and turbines. Waterwheels are clean but inefficient sources of power, and they require placing factories that need power close to sources of flowing water, which is not always possible or desirable. Steam engines are compact and more portable, but the coal they burn is a dirty pollutant, as we are by now all well aware. Oil is not much better. Thus, powering the first two industrial revolutions meant burning vast amounts of coal and oil—hence the environmental degradation.

Despite the problems they created, the technological advances made during the century and a quarter that covered IR1 and IR2 were quite astonishing and of huge importance for the lives of people. The industries they gave rise to and the downstream effects of these industries in terms of global ecology and international politics are with us today and their legacies, for better or worse, will be with us for quite some time yet. The big motor companies and oil companies were products of IR2, though their full effects on people, nations, and the world would not be felt until the late twentieth century. To appreciate the changes brought about by IR1 and IR2 I need hardly paint you a whole picture, but just present the broad brush strokes and one detail. The broad-brush effect is on living standards: preindustrial life consisted of stagnation—living standards of people across the world at the beginning of IR1 were pretty much the same as they had been two hundred years before. These living standards, taken as a whole, improved during IR1 and got better faster during IR2. As industrialization spread and continues to spread across the world, more and more people are living longer, healthier, more productive, and more comfortable lives than their preindustrial ancestors. One example of fine detail is to consider the small armaments of advanced countries at the beginning and end of IR2. At the beginning, say the mid-1800s, the standard army weapon consisted of a smooth-bore caplock (a simple modification of flintlock) musket. At the end of IR2, it consisted of powerful rifles with an effective range ten times that of the muskets, in addition to machine guns. It does not take too much historical investigation to conclude that industrialization has been, and will continue to be, something of a Pandora's box.

● ● ●

The third act of this long-running saga (long-running in human terms, but the merest blink of an eye geologically—the very start of the Anthropocene epoch) began around 1990, when the digital revolution began to have an impact on manufacturing. Digital electronics had been around for decades before 1990 and had even permeated popular culture—readers of a certain age will recall the digital games such as Space Invaders that seemed to take over every pub from the late 1970s. However, in these early years the digital electronics were confined to new products on the shelves and had not yet led to new ways of designing and making products, though the writing was on the wall. The driving force of this digital revolution has been the miniaturization of transistors onto integrated circuits, with consequential increases in computer processing speed and memory density, along with falling prices, as described in chapter 6. The third industrial revolution, IR3, is the rapid development and application of digital electronics, combined with the rapid expansion (and exploitation by industry) of the internet.

So the onset of IR3 is here taken to be 1990. A year earlier, the world wide web had been invented by Tim Berners-Lee, making access to information over the internet easy for everyone. Since then, the spread of information networks has increased, along with the ways people have found to use them. Thus as of 2015, some 42% of humanity had access to the internet—over 7 times the number just 5 years earlier.* The first digital cellphones arrived in the 1990s and seem to have taken over the world. The communications sector is just one part of industry that has converted to digital; others include computers, media and printing, retail, timekeeping, and navigation. Hence many people do not buy newspapers any more—they tap their tablet first thing in the morning; most people do not bother with watches anymore, even digital ones—timekeeping (and a camera) comes free with a smart phone; few people consult maps when it is easier to use a Tomtom GPS receiver—or a smart phone app.

*The figure is from www.internetworldstats.com.

The rate at which the world is changing now can be encapsulated in a paragraph by considering just the communications sector—then let us move on to the wider, deeper, and more profound ways in which the world is going to change in the near future as IR3 takes hold. So, faster and better digital communications between individuals is rapidly leading to the demise of letter mail, telegrams, typewriters, fax machines, landline phones, and pay phones. I recall as a graduate student in Scotland in the 1970s writing a letter to the United States, requesting a paper copy of some research, which was promptly snail-mailed to me—the process took *six weeks*. Thirty years later from my home in western Canada I emailed a request for a paper from its author in Australia, and got it sent as an email attachment within *two minutes*. Why buy encyclopedias when we have Wikipedia online? Why sit in on a traditional university lecture course when there are MOOCs* online? Why write a complaint to a newspaper when we can blog or post on Facebook or tweet to a much wider audience online?

So the combination of digital electronics and internet has changed the way we currently live and the things we make. It is now beginning to change the basis of national economies and will soon transform, if not end, industrial production as we know it in many developed countries. Already, automation is reducing the blue-collar workforce and increasing productivity per (remaining) worker. Some car makers are currently producing twice as many vehicles per worker as they did a decade ago. Automation reduces the importance of labor costs (a $499 first-generation iPad required only $33 of manufacturing labor). The new reality of manufacturing in the internet age—brought about by smarter and more dexterous robots, new materials, 3-D printing, and web-based services—is a changeover from mass production to mass customization of many types of manufactured goods. Mass production, a creation of IR2 that persists in many (usually outsourced—see chapter 10) industries today, requires much capital outlay to build large, dedicated factories in urban locations. It generates economies of scale, as we have seen. Automation and the internet will enable many products to be customized on small-scale assembly lines—bespoke production—as manufacturing moves back from factory to home. An entrepreneur with an idea, a laptop, and a 3-D printer will be able to run an international

*Massive open online courses—lectures put on the web for anyone to view.

business from home. Start-up costs will be much lower than for an old-fashioned factory; there will be no economies of scale, but instead small production runs for niche markets. Your hearing aid or eyeglasses or medical appliance, your customized automobile parts, specialized engine components, personalized gardening tools, or computer keyboards or door handles or food containers may be made in a neighboring country or a neighboring street, in a shack across the world or in a loft across town.

Here is the new business plan for many of your grandchildren: they design a gizmo that they feel the world needs. They float the idea on a crowdfunding website to raise cash. The idea catches on and they are sent the seed money to start their company—not only the money, but they have obtained market research and their first customers.* They print gizmos with a 3-D printer in their garage, obtaining raw materials from local sources found online. If they needed design and manufacturing expertise, that is also available online. If their gizmo proves to be popular, they increase production by outsourcing to an old-fashioned IR2 factory in China.

The IR3 world will very clearly be organized differently. Society will change as more people work from home and fewer commute, as the nature of work changes back from large companies to craft industries (not for all of industry, to be sure, but for many new industries that will arise and for much of what remains of the current manufacturing sector in developed nations). There will be challenges for governments as the line between manufacturing and services blurs. Copyright lawyers will have a field day. Unions will be squeezed even harder than they are now—there are no unionized robots. Totalitarian regimes will find it harder to suppress dissent, just as liberal countries will find it harder to suppress radicalization of its citizens by terrorists.† Mass surveillance of individuals and their communications will raise human

---

*This is exactly what happened to a bio-tech worker who thought the world needed a kind of lava-lamp filled with live jellyfish. Seriously—see Anderson (2012). He was hoping to raise $3,000 and received $130,000 within a month. Recall that the new IR3 craft industries will be small and so will require relatively small amounts of start-up money.

†Thus the Arab Spring revolutions were facilitated by social media, and the ISIS outsourcing of terrorism by radicalizing western Muslims is facilitated in the same way.

rights issues. Economic blocs and trade boundaries will become fuzzier. Pushback against IR3's social changes is now very evident in western democracies, with objections to international trade deals such as the Trans-Pacific Partnership, and the rise of politicians who oppose globalization, economic migration, and the outsourcing of jobs (e.g., Brexit in the UK and the election of Donald Trump in the US).

The next technical phase of the third industrial revolution is surely the much-anticipated *internet of things*. Machines will become partially self-aware via sensors; they will become intelligent and connected. Thus an airplane or any complex machine will be able to tell us when it needs an overhaul or maintenance and what specifically it requires to continue doing its job and functioning safely—this soon-to-be-commonplace capability is termed *predictive maintenance*. Turbines on a wind farm will be able to talk to each other so they can adjust blade pitch, to collectively generate power efficiently. Cars will drive their owners to work, and not the other way around. Connected sensors will monitor and assess activity across a city, aiding security and safety. Here are a few more examples of the interconnectedness between people and things—and between things and things—that will increase functionality or generate emergent behavior or just make life easier. Devices, as well as people, will be able to access the internet; smart buildings will optimize heating and ventilation in an energy-efficient way; smart dust (literally, micro-computers) will be injected into bloodstreams and blown into air vents to diagnose and to measure chemicals; information will be transmitted via streetlights; your office equipment will reorder supplies; your smartwatch will inform your doctor of your changing heart condition and blood pressure.

If, despite the last couple of paragraphs, this description of IR3 is sounding a little too rosy—too nice, like a serenade—then let me insert here an off note, by way of antidote. The last two industrial revolutions most certainly had their down side, in terms of human well-being. Here are a couple of chilling examples from IR3 that may serve as a warning about the future.

It is now possible to print a handgun. That is, downloading the design of a handgun to a 3-D printer, a functioning gun can be reproduced. It has also been recently demonstrated that a handgun can be fired remotely from a small quad-copter (a small multirotor helicopter, usually unmanned). I cannot think of a benefit to society of either of these developments. How long before we have our first fly-by shooting? How will the perpetrator be traced? These grim examples (here's another: flying IEDs [improvised explosive devices], i.e., terrorist bombs delivered by quad-copter) do not point to IR3 as bad in itself, but to a negative consequence of it due to human nature, or rather to the nature of some humans.*

 ## 6 Moore's Law

Any list of luminaries from the industrial revolutions must include the likes of Isambard Kingdom Brunel from IR1, Henry Ford from IR2, and Gordon Moore from the initial phase of IR3. Moore was cofounder of microchip giant Intel, and in a 1965 magazine article he proposed what eventually came to be known as Moore's Law. The initial formulation of this "law" (really an empirical observation) stated that the number of component parts on an integrated circuit (a microchip) was doubling every year. That is, fundamental components such as transistors were shrinking in size, and consequently digital memory and microprocessors were becoming smaller and the speed or *clock frequency* of the chips was increasing. By 1975 Moore amended his law to say that components were doubling every two years, to reflect a slowing from that initial spurt following the development of integrated circuits. This second formulation has stayed with us for four decades.

The effects of Moore's Law on the performance of digital electronic devices is astonishing. Memory expanded, yet devices shrank; algorithms grew in size and sophistication; the resolution of digital images was increasing as pixel counts skyrocketed. Intel CEO Brian Krzanich provided a telling analogy by asking us to consider a 1971 VW Beetle: if this car had advanced at the pace of Moore's Law over

---

*The future IR3 is described by Anderson (2012, 2014), the *Economist* (2012), and Rifkin (2013, 2015). For the future impact of 3-D printing, see Hammes (2015); for more on 3-D printed handguns and handguns fired from quad-copters, see Bilton (2014), Corcoran and Connors (2015), Gibbs (2015b) and Greenberg (2014); for the threat of flying IEDs, see Franke (2016). For the *internet of things* see the internet or Ferber (2013). Some worrying consequences of the internet of things for global security have been pointed out by former CIA Director John Brennan (2016).

the next 34 years then "you would be able to go with that car 300,000 miles per hour. You would get two million miles per gallon of gas, and all that for the mere cost of four cents" (Sneed, 2015).

In 1990 the size of an integrated circuit component was 800 nm (nanometers—a nanometer is a billionth of a meter); by 2012 it was down to 22 nm.* Of course, this reduction in size cannot continue forever. There is a hard limit when transistors are reduced to the size of a single molecule, which will happen soon. The demise of Moore's Law has been anticipated for a while, and yet it keeps going (more or less; Intel has admitted that today the doubling time is closer to 2½ years). Moore himself said in 2005 that his law would fail in 10 or 20 years time—Moore no more, as it were.

The law became very well known to the extent that, from an initial *describer* of microchip industry progress, it turned into a *driver* of progress. Intel and other microchip manufacturers tried hard to keep the law valid: chip density will double every two years—to this extent, Moore's Law has been something of a self-fulfilling prophecy. Progress over the decades was not at a uniform pace, of course; at different times over the past 50 years key inventions or developments have been made with lasers, DRAM (dynamic random-access memory), flash memory, metal-oxide semiconductors, photolithography . . . the list is a long one. When Moore's Law finally fails, we will have pushed this particular technology to its limit, and further progress will have to come from a quite different direction. It is possible that the coming failure of Moore's Law may define the end of IR3, but I wouldn't bet on it.[†]

*In 2000, the clock frequency of a typical digital device such as a laptop was 1.3 gigahertz (GHz), which had increased only by a factor of two to 2.8 GHz by 2009. However, the laptops of 2009 had quadcore processors, whereas those of 2000 were single core—thus the effective processing power of the 2009 laptop was about eight times greater than that of the 2000 machines.
[†]A lot of electronic ink has been spilled recently defining and discussing Moore's Law. See, e.g., Bradshaw (2015), Clark (2015), Gibbs (2015a), and Sneed (2015).

⑦ Building BRICS

The acronym BRICS (Brazil, Russia, India, China, South Africa) is one of many that stand for different groups of emerging nations, meaning nations with economies that are rising out of preindustrial agriculture and are developing industries, with all that implies (infrastructure, burgeoning and urbanizing population, rising living standards, increasing pollution, greater power consumption, greater influence in the world). Other acronyms include MINT (for Mexico, Indonesia, Nigeria, Turkey), EAGLE (Emerging and Growth-Leading Economies), NIC (Newly Industrialized Country), and N-11 (Next 11). The group of Brazil, Russia, India, and China (BRIC) leads the list in terms of development; it was formed in 2001 and expanded to include South Africa in 2010.*

The BRICS nations between them have 42% of the world's population, as of 2015, and 20% of the gross world product—but their collective economy is set to dominate that of any other block before we reach the year 2050 (though at the time of writing, most of the member nations are stalling, economically; Brazil and Russia are in deep recession, and China is having problems switching from export-led growth to domestic consumer growth led by its burgeoning and restive middle class). The five nations have grouped together for mutual benefit. They are setting up the New Development Bank, which they consider will better serve their needs than do the European- and American-dominated IMF (International Monetary Fund) and World Bank. Joining will also give them more political leverage on the world stage; to this end the BRICS nations meet annually to discuss internal and external policy.

South Africa was probably included to give the group a presence in Africa. China and India are industrializing fast (China is in front, with five times the GDP per capita of India). The situations and problems of the member nations vary; they share a desire for economic growth and development, and are less interested in human rights or

---

*Much of this summary of the BRICS nations and their prospects comes from Blackhurst (2015), Desai and Vreeland (2014), Kupchan (2012), Naude et al. (2015), Sharma (2012), and the BRICS Information Centre website at www.brics.utoronto.ca.

environmental issues than are the more highly industrialized nations. China seems to be moving upscale by growing more and more hi-tech industries, such as aerospace and telecomms; India may be positioning itself for moving into China's shoes, taking up the large-scale manufacturing of less expensive goods. Brazil and Russia both have extensive natural resources. Other than their economic ambitions, the BRICS nations have little in common in terms of culture or political heritage.

An interesting feature of economies that are industrializing today, compared with those that developed industries as part of the first or second industrial revolutions, is that they can leapfrog: go from imitation of mature technology to innovation in the same technology over the course of a few years. They can skip stages: go from a rural agrarian economy to IR3 in one step, without bothering with steam engines or trade unions, without the dead ends and blind alleys of cutting-edge innovation—they do not need to reinvent the Bessamer process or rediscover electrification. This fact leads to odd disparities. For example, many of the poorly developed nations in the world today have widespread use of cellphones but few roads. Business is being conducted by cellphone, which is driving economic development rather than being a product of it. BRICS nations are considerably further down the development road but still have elements of all three industrial revolutions within their economies. Thus, Russia has a leading role in space and in the aerospace industry, but with institutional corruption at Third World levels (recall that efficient infrastructures such as banking are deemed a necessary precursor for industrialization—see chapter 5). We find steam trains in India alongside international call centers; we find huge factories burning coal in China along with leading-edge solar power development. (Given the severe pollution in China—a direct result of their IR2 industries—about the only "blue sky" you will find in Chinese cities is their research.)

And here is the problem. The pollution that was generated in Britain, the United States, Germany, and the other industrialized nations during IR1 and IR2 is now being generated by BRICS nations and those of all the other "acronym economies" that are gearing up their industries. The climate consequences of nineteenth-century polluting industries are still with us, and the addition of much larger current sources of pollution around the world may lead to pollution levels that endanger human health on a global scale, perhaps existentially if they drive uncontrollable climate change. Hence the extensive international emission control talks and (so far mostly toothless) protocols for curbing the pollution that is generated mainly by industrializing nations (see chapter **19**).

Peppered Moths

Found across much of the northern hemisphere, these little moths have a speckled appearance, most likely for camouflage. The overall coloration varies from pale tan to black; this variation of hue applies in ultraviolet light as well as in visible light, in the following sense. A pale peppered moth resting on a pale background is harder to see in both visible and ultraviolet light than is a pale moth against a dark background. Birds are the main predators of these moths, and most birds can see ultraviolet as well as visible light, hence the value of this broadband camouflage. Peppered moths rest against tree boughs, so there is a survival advantage for those moths with wing patterns that closely match the color of the tree on which they are resting. Evolutionists say that there is *selection pressure* for the moths to adopt a very similar color to the trees they commonly rest on, because those moths are less likely to be picked off by predators and so will be more likely to reproduce and pass on their genes, and thus their coloration.

At the beginning of the nineteenth century, the north of England was being transformed by IR1 into a bleak landscape of dark satanic mills belching smoke, of land blighted by industrialization and of trees—here is the connection with moths—blackened by soot. Naturalists noticed that the peppered moths around the growing industrial city of Manchester were noticeably darker than they had been a few decades earlier; when Darwin published his theory of evolution a generation later, this example of peppered moth coloration change was taken to be a good illustration of natural selection. By the middle of the twentieth century, pollution control legislation and changing industrial practices eliminated most of the soot, and so the tree boughs around Manchester returned to a paler color—and the moths followed suit shortly afterward.

Confirmation of Darwin's theory, you might think, and I would agree with you. However, it seems that some of the experimental techniques used to test the notion of adaptation of color of peppered moths were open to question. A few scientists called the experiments into doubt, and some creationists called the whole theory of evolution into doubt as a result. Later, the experiments were performed more rigorously, convincing the scientists, though presumably not the creationists, that the case of the peppered moth coloration in the north of England is indeed a very good example of Darwinian natural selection in action.*

As well as telling us a lot about how quickly some forms of life can adapt to manmade environmental change, this example tells us something about how industrialization has led to environmental degradation, right from the early days of the first industrial revolution. Of course, the story does not stop there.

• • •

The effects of industrial pollution have spread beyond the north of England in the 1800s, and beyond peppered moths. Most of the developed world has suffered the effects for at least a century, until more-or-less effective pollution control legislation came into being in the late twentieth century to reduce the more obvious sources; most of the developing world is suffering from the negative impacts of pollution now, as they lack the will to implement or the power to enforce environmental legislation (of which, more in chapter 19).

Pollution of rivers and canals, and of lakes and oceans, happened directly from the factories as waste from industrial production was poured into waterways. Untreated sewage from the rapidly growing urban populations was tipped into rivers and then, when this short-sighted practice—clearly the practitioners could not see beyond the end of their noses—led to a stink, was carried out to sea and dumped. Water became polluted indirectly when newly invented fertilizers and pesticides formed part of the runoff of stormwater into rivers and oceans.

---

*I have read a retraction from one of the doubting scientists, but have read no such retraction from any of the creationists; hence my presumption. For technical and popular accounts of peppered moth coloration adaptation, called *industrial melanism* in the literature, see Cook et al. (2012), Majerus (2009), Marren (2009), Webb (2016), and Zimmer (2013). See also the Wikipedia website on the subject.

Oil spills are the most obvious, most visible, and therefore most reported examples of water pollution in the world today. This visibility has led to effective legislation in western countries, and consequently the pollution of rivers and oceans with oil, plastic, fertilizers, pesticides, and other chemicals is not the worst of our own current pollution concerns, though every time a large oil spill occurs, it becomes the most pressing.

The worst pollution that is happening in the world today—by far, in terms of human health—is air pollution. In the short term, at least, it is a greater risk to humans than is global warming. According to the World Health Organization (WHO), one in eight of the total number of human deaths each year is due to air pollution in one form or another. That figure amounts to seven million people dying prematurely each year around the globe as a result of cardiovascular diseases, respiratory diseases, and cancers—deaths that would not have occurred if the air that these people breathed had not been polluted. The distribution of these deaths around the globe and within society is not uniform: low- and middle-income people in Southeast Asia and the western Pacific are worse affected than average; people in North America are less affected than average—but see below. A recent study published in the prestigious journal *Nature* gave a figure of 3.3 million premature deaths per year worldwide (mostly in Asia) due to outdoor air pollution. It is not clear whether this figure is consistent with the WHO figure, but it is alarming enough: HIV/AIDS and malaria between them kill fewer people each year, and the figure of 3.3 million will double by 2050 if the world does nothing to reduce air pollution between now and then.*

In some cases, pollution of the atmosphere is as obvious as that of the water and land: for example, the air in Chinese industrial cities is famously and visibly polluted—16 of the 20 most polluted cities of the world in 2007 were in China (although by 2013, the majority of the 20 most polluted cities were in rapidly industrializing India). The level of pollution may be less intense than that in the vicinity of tailings ponds downstream of Canadian tar sands, but far more people are affected. There are occasional visible disasters such as happened in Bhopal, In-

---

*For the air pollution figure, see Brink (2015), Caiazzo et al. (2013), Lelieveld (2015), and Ubelacker (2013).

dia, in 1984, when 30 tons of lethal methyl isocyanate gas was acciden-
tally discharged into the atmosphere from a Union Carbide plant:
between 2,500 and 8,000 people died. In London there were around
4,000 extra deaths (i.e., deaths above the normal background level) in
December 1952, when a meteorological inversion layer trapped smoke
from millions of residential chimneys and thousands of industries to
form a toxic smog. This event led directly to clean air legislation in
Britain, in 1956. A similarly visible disaster in the United States—in
Donora, Pennsylvania, in 1948—led to the Clean Air Act of 1963 and
1970. Again, it was a combination of industrial pollution (from a zinc
smelting plant) and a weather anomaly that trapped smog over Donora,
killing 25 and hospitalizing 7,000 with respiratory problems.

The vast majority of deaths due to air pollution are not nearly as
visible, and so the sources of the pollution that lead to these fatalities
are not under as much pressure to reduce emissions. They are invisible to us not just because they occur overseas in developing countries,
but because they are literally invisible; we all receive a low dosage of
air pollution every day, year in, year out, from industrial smokestacks,
power plants, automobile tailpipes, rail and sea transport, and domestic heating systems. An MIT study shows that this air pollution leads
to 200,000 premature deaths each year in the United States. The worst
offenders are automobile exhausts (carbon monoxide, carbon dioxide)
and electrical power generators (ditto, plus sulfur dioxide and nitrogen
dioxide); the worst-affected state is California, and the worst-affected
city is Fresno-Madera, California, according to the American Lung
Association website.[*]

Industries pollute our air with gases, with lead and other metals,
and with particulate matter. The carbon dioxide leads to global warming (see chapter 16) while the particulate matter (especially the fine
particles that are less than 2.5 micrometers in diameter) lead to heart
diseases and respiratory ailments.[†]

---

[*]See the American Lung Association website, http://www.stateoftheair.org/2015
/city-rankings/most-polluted-cities.html.
[†]For pollution figures, see Duggan (2015), Ellis (2013), Lagorio (2007), Park
(2014), Walsh (2013), and the WHO news release "7 million premature deaths annually linked to air pollution," http://www.who.int/mediacentre/news/releases/2014
/air-pollution/en/ (March 25, 2014).

On the whole, humans have benefited from the industrial revolutions past and present, at the cost of polluting our environment. Globally, pollution is a problem that is killing millions, and the problem is likely to get worse before it gets better, as more and more nations develop industries before they develop the will or the means to avoid polluting their land, water, and air. But pollution is not the main issue as we head toward the Anthropocene—it is a serious problem now but one that may be solved or at least mitigated in the decades and centuries to come. We know how to fix it. An encouraging example of fixing a specific pollution problem is the way we have solved the problems caused by CFCs—see chapter **21**.

 Globalization

A much-used and somewhat misunderstood word, globalization is sometimes considered to mean little more than the outsourcing of manufacturing—sending jobs abroad to developing nations. Perhaps it also includes importing tomatoes from Mexico to Canada or lettuces from Jordan to England.* Globalization certainly includes these kinds of things, under the umbrella "economic integration," meaning the integration of national economies (see chapter **10** for a telling example of smartphone fabrication), but it is much broader than that. Globalization is the rapidly expanding interconnection of peoples around the world via digital communications, computer technology, and transportation; it is the integration of ideas, cultures, laws, and social movements, as well as economies. In the long run it can be seen as bringing about the decline of nation-states as we know them.

---

*Both of these transactions occur in real life. Here is a yet more global example, from a study about the sourcing of Irish food (see Safefood, 2009). The ingredients for a pizza made in Ireland—for local consumption—originate in the following countries: wheat (from one or more of Belgium, France, Ireland, India, Sweden, UK, US); processed cheese (Belgium, Germany, Ireland, UK, US); processed tomatoes (Australia, Belgium, China, France, UK); ham (Brazil, Chile, France, Germany, Japan, UK); pineapple (Costa Rica, France, Guatemala, Panama, South Africa, Spain).

Improved communications and transport means that the world is shrinking: it is much faster to send legal documents, photographs, technical data, books, and other such information across the world today than it was a generation ago (from the speed of an airplane to the speed of light), and it takes significantly less time and money to transport people and the goods they trade. Trade restrictions between countries were reduced with the GATT (General Agreement on Tariffs and Trade) treaty, which ran from 1947 until 1994, when it was replaced by multilateral trade agreements organized by the World Trade Organization. Globalization facilitates the movement of people and things by easing border controls through "open skies" policies (many low-cost airlines sprang up in the 1980s) and free trade zones such as those set up by NAFTA (the North American Free Trade Agreement) and the European Union. To ease the flow of goods, capital and labor markets are being deregulated. It is easy to see how economic integration has financial and political ramifications.

The second industrial revolution led to bulk trade and movement of peoples across continents via railroads, and across oceans via freighters. Refrigerated trucks and rail cars, bulk carriers, and container ships—all bring goods from across the world to our supermarkets and big box stores, which themselves are no longer local companies that can readily be assigned a nationality. Multinational corporations emerged, and are now morphing into something else as new business models change the game (see chapter 10). The benefits of economic globalization are lower cross-border transaction costs, including labor costs and the costs of raw materials, and new business opportunities. The result for an increasing number of people—those with purchasing power in all but the poorest countries—is reduced costs for all kinds of goods.

Cultural integration has been around for centuries, if we include the spreading of religions across borders and continents. Modern cultural integration is more rapid and widespread, if more frivolous. McDonald's and Starbucks might almost be used as a metric for globalization, so far and wide have they spread over the globe. Music such as jazz, reggae, and rock has disseminated from its western origins. Much but by no means all of cultural globalization has been from the West to the rest of the world; English is the lingua franca of the internet and so of globalization.* A superficial look at cultural and social globalization has led some people to call it "Americanization," but this label is misleading. Yes, there is a McDonald's in Panama City (at the time of writing there are three in the center of the city) but there are also over 120,000 Panamanians in the United States, some of whom no doubt work at McDonald's. Your T-shirt or baseball cap may have an American flag on it, but more than likely, it was made in China. A utility company in the Midwest may interface with its local customers via a call center in India; an American hi-tech company outsourcing some of its software to India may very well have Indian managers in the United States.

Globalization is uneven; it proceeds in spurts, and the degree to which it is evident depends on whereabouts you are in the world—but it is spreading wider and deeper everywhere. Not all aspects of globalization are good, of course. Arguably, the dissemination of the Arab Spring via social networks was a good thing, but surely the recruitment of ISIS supporters via the same networks is not. Outsourcing fabrication lowers prices in stores but puts pressure on local jobs and wages. Pollution is exported to countries where it is less controlled. There are worries about the safety of imported products (food, pet chow, and children's toys from China, for example). Inequality around the world, including within the United States, is increasing. A financial crisis originating in the United States is exported to the rest of the world. In the week I write these words, worries about the Chinese economy are causing jitters in global stock markets.

Vastly improved communications compared with those available half a century ago cuts two ways. Yes, we can trade and get to know each other better, but people being what they are, it also leads to problems and potential instability. A cartoon in a Danish newspaper or a French satirical magazine creates outrage in the Moslem world and leads to acts of terrorism. Cellphones facilitate communications between peoples, but also set off car bombs. Improved communications means easier transfer of money across borders, but also currency wars. Neither finance nor economics are well understood as disciplines of academic study, let alone as drivers of industry out in the street (see chapters 34 and 35), and the instability of global finance and global

---

*Appropriately, because English as a language shows diagnostic signs of globalization in its many foreign loan words and phrases, such as *lingua franca*.

economics are therefore *increased* by the interconnectedness that is globalization. Think of Saudi policy in Yemen influencing world oil prices, the politicization of global trade talks because of local disputes and intensified competition, the ease with which economic sanctions and sanctions-busting can be organized. A fire in a Chinese industrial city or a Bangladeshi sweatshop can raise prices in Glasgow or lead to demonstrations in Ottawa. Mortgage debt repackaging in New York can lead to pension fund collapses in Oslo, or anywhere.*

The undoubted economic and social benefits of globalization must be set against the economic, environmental, social, financial, and political risks. Consider this: expanding the European Union has caused increased economic and financial instability—exactly the opposite of its primary goal—in part because of structural flaws, but in part because the number of sources of fiscal incontinence and economic irresponsibility has increased. Consider also the recent populist backlash—due in part to globalization, and its immediate consequence, the free movement of people across borders. "Brexit" and the rise of Donald Trump can be attributed to this backlash, often dubbed "populism" and a feature of the political left as well as the right. Here is what Klaus Schwab, founder of the World Economic Forum, says about globalization: "The speed of the current breakthroughs has no historical precedent. When compared with previous industrial revolutions, the Fourth is evolving at an exponential rather than a linear pace. Moreover, it is disrupting almost every industry in every country. And the breadth and depth of these changes herald the transformation of entire systems of production, management and governance." IR4? I didn't get that far in chapter 5, but many people are now classifying globalization and the new technology that leads to it as an industrial revolution distinct from IR3, such is its significance.†

---

*For interesting insights into different aspects of globalization, see, e.g., Bauman (2000), Bremmer (2015), Hammes (2016), Mourdoukoutas (2015), Smith (2014), and Stiglitz (2007).

†The recent backlash against the economic effects of globalization (I should say, the backlash that has recently become very obvious politically, across the Western world) is discussed by Blyth (2016), Gollom (2016), and Zakaria (2016).

10 Smartphones Are (from) Everywhere

These ubiquitous, essential, and annoying machines are an elegant and apposite icon of modern hi-tech manufacturing and business models. The economic worlds of trade and production are not what they were when you were a kid. As recently as 20 years ago it made sense to speak of or rant against multinational corporations, but today they don't exist, at least in the sense they did back then in industries such as semiconductors. The business world was full of technical terms then as now—that hasn't changed—and the large chip manufacturers that arose as a consequence of the digital electronics revolution were called IDMs, for Integrated Device Manufacturers. A company such as Intel bought raw materials, designed integrated circuits, manufactured the microchips, and then sold them. They were integrated vertically—they controlled every stage of chip fabrication from a few sites near their headquarters in California. But the Intel that made Pentium chips for your desktop computer in the 1990s is not the same company, except in name, as the Intel that makes microprocessors for smartphones today. The same is true of all major companies in the semiconductor industry and many other industries across the world.

Let's look at smartphones, as an example. The raw materials that make up smartphones are very diverse. Of the 90 or so elements that exist in nature (plus a couple of dozen that are so unstable they do not exist in nature, but can be created in nuclear reactors), over 60 are required to make smartphones. Twenty-five percent of the weight of your smartphone is due to silicon, the ubiquitous semiconductor; the next most common element is iron (20%). From there the metals aluminum, copper, lead, zinc, tin, nickel, and barium compete with plastics (hydrocarbons—combinations of hydrogen and carbon). Between them, these elements contribute 93% of the weight of a typical smartphone. The other 7% comes from 50 elements, including these rare earths: cerium, dysprosium, europium, gadolinium, lanthanum, neodymium, praseodymium, terbium, and yttrium—you would be forgiven for having never heard of them. But smartphone fabricators have heard of them, and they know of their unique properties, without which smartphones would not work as well as they do.

The organization of smartphone manufacturing* is determined partly by geology (for example, some 90% of rare earth elements are mined in China) but mostly by today's globalized business model. Smartphones are very complex machines with hundreds of components, each the result of complex construction processes; they are mined, processed, fabricated, manufactured, and distributed worldwide. Thus there are about 340 iPhone *fabs* (fabrications—in the sense of construction rather than deceit) in China alone, 150 in Japan, 75 in the United States, 30 or 40 in South Korea and in Taiwan (the exact numbers for each nation depend on whom you consult), 20 or so in Malaysia, in the Philippines, in Thailand, and in Singapore, and between 1 and 10 in Mexico and Latin America, in the United Kingdom and Ireland, in Germany and Austria, Belgium and the Netherlands, France and Italy, the Czech Republic and Hungary, Israel, Vietnam, and probably one or two other countries that I have overlooked. Other types of smartphone may have a different mix of material and fabrication origins (Moto Xs are largely fabricated in the United States, for example), but you get the idea: literally hundreds of companies make components for the final product, from all across the globe.

Modern communications and cargo ships make all this interconnectedness—of manufacturing and trade as well as digital electronics—possible. These developments have changed industry from the days of monopoly IDMs. Such large companies have adapted to the changes—indeed have driven them—and the name of the game is *specialization*. Intel, which used to make everything digital from memory to microprocessors, now specializes in the latter. Samsung has arisen to dominate the memory market. Japan is still good for cameras. Texas Instruments has specialized, from general purpose calculators to cellphone chips. Manufacturing is dominated by China. Design is still largely concentrated in the United States (with Taiwan joining in), while new chip fabricators—known as *foundries*, presumably because they make the basic units—are concentrated in Asia. There is no vertical integration;

---

*I am using the words "manufacture" and "fabricate" as synonyms, but in industry these words have slightly different meanings. Roughly, components are fabricated and the finished product is then manufactured from them.

one Asian foundry will provide microchips to many different smartphone or desktop or tablet manufacturers. The foundries serve *fabless* firms, i.e., companies that do not make anything themselves, but instead outsource (a modern buzzword) all fabrication and manufacturing to others. The reason for this specialization is largely the different management cultures that are needed for different types of work; thus, for example, the skills and expertise required for managing a design company are very different from those needed at a fabricator. The semiconductor industry could nowadays be more accurately described as horizontally integrated instead of vertically integrated like the old IDMs, and it is spread across the planet rather than being concentrated in one area, such as Silicon Valley.*

Specialization and international outsourcing are the hallmarks of globalization. One of the main driving forces for globalization is cheap labor abroad; one of the main benefits is lower prices; one of the main risks is the rapid and far-reaching spread of economic downturns, exacerbated by ever-more-complicated geopolitics.

● ● ●

The industrial revolutions have led to unprecedented and increasingly rapid changes for humankind. The quality and length of affected human lives has risen hugely over the last 200 years. The organization of societies and the interactions between nations have changed. Every place is interconnected. These societal changes have benefited people, in general. However there are two ongoing and very bad consequences of industrialization and globalization: industrial pollution and a sixth mass extinction. A further, knock-on effect is global warming—potentially more serious for us than a mass extinction and longer lasting than industrial pollution.

*There is a large body of literature out there about the evolving semiconductor industry. For a nontechnical introduction, see Cuynar (2006) and the online presentation of Hodges and Leachman. For smartphone raw materials data, see Ashton (2013), www.mobileburn.com (2014), and Rathi (2013). In researching this section, I came across the following quote from Ken Beyer, CEO of an electronics recycling company: "There's more gold in a pound of electronics than a pound of gold ore." But this takes me into a whole other sector of the industry, which I will leave for you to investigate independently. From recovering and recycling, you can quickly move on to Third World exploitation; there's a website to help you calculate your slavery footprint, should you wish to know about it (see http://slaveryfootprint.org/).

It is now time to look into the world that our children and grandchildren will inherit. How many people will be here in the world, and what kind of climate will we hand down to them?

11 The Population Bomb

In 1798 the reverend Thomas Malthus provided us with a pessimistic prediction for the future of our species, based on population growth. He said that the number of people in the world was going to increase geometrically (meaning exponentially), whereas their food supply would grow only arithmetically (linearly) and so at some point there would be more people than the world was capable of feeding. Result? The world population was heading for a crash, due to starvation. This Malthusian notion has remained in the public consciousness ever since it first appeared. There have been intermittent food shortages and localized mass starvation, but not yet on the scale that Malthus envisaged. An update on this disturbing theme occurred in 1968 with the publication of Paul Ehrlich's *The Population Bomb*, which foresaw mass starvation in the 1970s and 1980s due to the population increases of IR2. It didn't happen—again—because of the green revolution, which brought about an unprecedented increase in crop planting and productivity (see chapter **12**). Yet Ehrlich persists with his belief that we are about to hit the skids due to overconsumption of resources.*

Here are the basic facts about the human population and its recent growth. As the first industrial revolution was getting into its stride, around 1800, the number of living human beings was about one billion. On August 23, 2016, it topped 7,445,413,000; it will be a little higher by the time this book goes to press. By the end of this century the number will be somewhere between 9 and 11 billion, depending on which population model you believe. The growth rate peaked at

---

*See Kolbert (2014) for an appraisal of Ehrlich's view. The numbers and projections for this section have been taken from Carrington (2014), Dimick (2014), Dusheck (2014), Sullivan (2013), the CIA *World Factbook 2016*, and an entertaining and enlightening Hans Rosling TED talk available on YouTube (https://www.ted.com /talks/hans_rosling_on_global_population_growth).

2.20% per annum in 1962 and 1963 and has since dropped to half that number. Expressed in a different way, the population increases at a rate of 2.4 people per second. (If you find the fraction distressing, let us call the rate 12 people every 5 seconds.) The birth rate is 4.3 per second; the difference is due to mortality. This rapid rise has been due, of course, to improvements in living standards across the world, and in particular to medical and hygiene advances and to greatly improved food production.

Some 25.6% of people alive today are under the age of 14; of these, 52% are male. Some 8.5% of the world population is 65 or older; of these only 45% are male—females last longer. The median age is 29.7 years, and life expectancy is 68 years and 4 months. A little over half of us live in cities—a pointer to the spread of industrialization. Eighty-six percent of us are literate; two-thirds of the rest are female. The stated growth rate is a global average, and the local rates vary significantly with region; they are highest in sub-Saharan Africa, the Middle East, and South and Southeast Asia. The population growth rate is negative in eastern Europe due to low fertility and emigration; in southern Africa due to HIV-related mortality; and in Japan due to low fertility. India is likely to replace China as the most populous nation in the world around 2028. The population of Nigeria will overtake that of the United States around the middle of the century. By 2100, Southeast Asia and Africa will have the highest population densities in the world. Currently, there are 2.0 children born to each American woman, on average, which is almost neutral: 2.1 children per woman are required to maintain a population by births alone. In fact, the US population will probably remain stable if the current level of immigration is maintained for the remainder of this century. (This stability is less certain on the day I write these words, April 23, 2017, given the immigration policy that President Trump is attempting to implement.)

Life expectancy will increase from the current world average of 69 years to 76 by 2050 and 82 by 2100. In the developed world, this figure will be 89 years.

The number of people in the world who are of working age (defined as 15–65) peaked as a percentage of the total global population in 2012. The fact that it is now decreasing will place an upward pressure on wages. This effect may dampen the rising inequality in the developed and developing world, but will not be welcomed by employers, who benefited in the past from a demographic "sweet spot" as baby boomers entered the workforce; they are now retiring and so are about to switch from being an economic driving force to being an economic drag. In these ways, demography influences global economics.*

There are aspects to demography—hinted at in the above numbers—that make the prediction of populations more difficult than you might think from the Malthusian geometrical growth idea. People live longer as economic conditions improve and infant mortality drops. Migration affects the population of a country, and can influence global population growth if migrant living standards change. There is a common phenomenon of developing countries known as the *demographic transition,* whereby such countries move from high to low birth rates and death rates as living standards increase. One unexpected and somewhat counter-intuitive consequence of this phenomenon is that, to reduce the birth rate of a region, it is necessary to *increase* the child survival rate.

The paradigm for world population development over the last 20 years has been this: population will increase to a maximum of about 9.5 billion by mid-century and will thereafter decline gradually to 9 billion by 2100. Recently, this paradigm has been challenged by a new model, which claims that the number of people in the world will increase to about 11 billion by the end of the century. The difference of two billion raises questions about food and water supply, waste disposal, energy supply, healthcare, and social cohesion. The different predictions show a large uncertainty—by a factor of two—about how the global population will change over the next eight decades. Yet accurate prediction is essential if we are to foretell where our climate is going, and if we are to plan meaningfully for our collective future. Thus, if we deem it necessary to limit the growth of humanity, past experience and demographic studies show that we need to reduce the birth rate in undeveloped countries by educating women and expanding family planning programs. Today, one person in eight has insufficient food; that fraction will increase unless we are able to limit population growth, increase food sources or more equitably distribute the food that is grown, increase available drinking water, and maintain a world climate that permits these changes.

*Weldon (2015b).

⑫ Manna from Science

The enormous population of the Indian subcontinent has always been vulnerable to food shortages and famines, largely due to the capricious nature of the annual monsoon. Historically, the number of people who died from malnutrition during a famine has been made worse (worse than the logistics of food availability and number of mouths to feed would suggest) by foreign occupation, human callousness, and incompetence. Since independence from Britain in 1947 there have been no mass famines but, even so, food shortages have led to thousands of deaths. The threat of famine did not go away with the arrival of political independence. Indeed, it was widely believed that India was heading toward a famine in the 1960s as a result of population increase, but fortunately it didn't happen. The burgeoning populations there and elsewhere in Asia have been mostly spared the specter of mass starvation as a result of the green revolution that began in 1960.

This start date is a little arbitrary; I choose it because 1960 is the year the International Rice Research Institute (IRRI) was formed—this institution figures prominently in the green revolution. I could have begun an account of this technological revolution in cereal crop production earlier and on a different continent. In the 1950s, Mexico sought to become self-sufficient in food, and did so with the help of the United Nations, the US government, the Food and Agriculture Organization, and the Rockefeller Foundation—a philanthropic organization. By injecting funding, developing irrigation, and introducing new crops and new breeds of old crops, this disparate group showed Mexican farmers and the rest of the world how modern scientific methods, organization, and cash could boost food production.

In the early 1960s, an American biologist called Norman Borlaug developed a semidwarf strain of wheat—he is credited with the amazing achievement of saving a billion people from starvation as a consequence, and in 1970 received a Nobel Peace Prize for this work. The IRRI was founded in the Philippines in 1960 by two American charitable foundations: the Ford Foundation and the Rockefeller Foundation. Within three years, a new high-yield rice variety, a dwarf strain that was the result of crossbreeding by plant pathologist Peter Jennings, had been developed. Applied to the subcontinent under Indian geneticist M. S. Swaminathan, these new strains of wheat and rice boosted production of the world's two most important cereal crops. Ten million tons of wheat had been produced in India in 1960; this number grew and grew—by 2006 the yield was 73 million tons. Indian rice yields increased from 2 tons per hectare in the 1960s to 6 tons per hectare by the 1990s, so that India is now a major exporter of rice.*

Not only India benefited. Cereal production (wheat, rice, and corn) across the world has more than doubled over the 56 years from 1960. This is the green revolution, and it saved the developing world from disaster. The United States had shipped millions of tons of wheat to Asia following World War II to head off mass starvation there; China suffered famine following Chairman Mao's Great Leap Forward; India, as we have seen, was on the brink. (Africa missed out on the benefits of the green revolution due to a combination of drought, corruption, and lack of infrastructure, which is why the mass famines that have occurred within living memory of most people are confined to that troubled continent.)

Money, science, irrigation, modern management, international collaboration, and information sharing: these are the inputs that led to this revolution in food production. Apart from selective breeding programs, the science of the green revolution involved developing, producing, and distributing fertilizers and pesticides. This was the era of DDT and monocultures, as well as of improved yields. The energy needed to produce the green revolution—to power farm machinery, generate fertilizers, and so on—grew faster than the crop yields, so that the food generated per unit of energy consumed actually fell. This makes the green revolution a dependent child of the oil industry; when oil dries up, the progress made in food production will be under threat.

Brazil massively expanded its soybean production as a result of the green revolution. Rice yields worldwide increased from 1.9 tons per

---

*The strain of rice that proved so successful is named IR8—it was the eighth cross that Jennings developed, and it is now called India Rice because of the massive impact it had on that country. It resulted from an international breeding program that is an early indication of globalization: a variety of rice from Indonesia was crossbred with a dwarf variety from Taiwan, following a search across the rice paddies of the world for suitable strains. The resultant IR8 is now grown everywhere that rice is cultivated.

hectare over the period 1950–1964 to 3.5 tons per hectare in 1985–1998. These and other gains led to an average annual global food production increase of 1.75% for half a century. This impressive increase was only just enough to keep pace with the population growth over the same period, and the rate of improvements (as new grains continue to be developed and irrigation, fertilizers, and pesticides are improved and distributed) have leveled off over the last 20 years. World grain production per capita peaked in the 1980s; grain stocks have never exceeded a few months' supply. Recent years have seen food riots in Bolivia, Cameroon, Egypt, Haiti, Indonesia, Ivory Coast, Mauritania, Mozambique, Senegal, Uzbekistan, and Yemen. A second green revolution is needed soon, or the disasters that were avoided by the first will happen. Some authors have likened the food situation in the world today to that of Ireland just before its famine in the 1840s. We are too many people who are too dependent on a few key crops (one, in the case of Ireland—the potato).

In fact, the second revolution in food production began a decade ago, and it is taking a decidedly modern, hi-tech turn. Genetically modified (GM) crops will cut across many species of food plant, not just the two or three main crops of the first green revolution. The changes will result in seeds that are customized to specific locations and specific issues. For example, a "brown streak virus" threatens cassava (a.k.a. manioc, tapioca), which is the world's third most important crop; work is underway to fix this problem by genetically modifying cassava. The richest rice-growing regions of Asia are the big river deltas such as the Mekong, Brahmaputra, and Irrawaddy; these are vulnerable to flooding with salt water as sea levels rise. Salt water kills rice, but work is under way to develop rice strains that are tolerant of salt. Even freshwater flooding kills rice; however, a strain that can remain under fresh water for two weeks has already been made (called IR64) and is being planted in regions that are prone to flooding, such as the state of Uttar Pradesh in India. New GM rice strains will require less water and nitrogen; nutritional quality—not just calories—will increase. The second green revolution is expected to benefit the most marginalized regions and peoples (unlike the first, which had the effect of exacerbating social divisions and increasing the gap between rich and poor in developing countries).

Some of the components of the first green revolution—monocultures, fertilizers, and pesticides—are anathema to some environmentalists.* Their objections, criticisms, and actions to oppose these features have led Norman Borlaug to criticize some lobbyists as elitists who have never suffered or seen the effects of hunger. The second green revolution will be centered on GM crops, already grown on 11% of the world's arable land and fully half of US cropland. Genetic modification is regarded with great suspicion by many environmental groups for several reasons, especially in Europe and Africa.† Some activists have trampled IRRI test plots. The powerful environmental lobby in the western world has caused cutbacks to green revolution funding that may be partly responsible for the leveling off of crop yield gains over recent years. IRRI director Bob Ziegler says he feels betrayed by the environmental movement. At the same time, many westerners squirm in their armchairs when they hear that Monsanto is developing "Roundup Ready" crops that are tolerant to that potent pesticide. On the other side of the world starvation looms because, as Princeton climate scientist Michael Oppenheimer says, the slowdown in the growth rate of crop yields over the last two decades is "the biggest threat of climate change."‡

Does it trouble you that some environmental activists are at odds with the green revolution? Feeding the growing number of humans necessitates a technological and political solution; the flip side of this solu-

---

*Monocultures increase susceptibility to pathogens. The dependence on non-renewables such as oil for energy and phosphorus for fertilizers creates a potential for collapse. (On the other hand, burning fossil fuels to grow crops does not increase greenhouse gas emissions, because the crops fix carbon.) Another criticism of the first green revolution is that its economic benefits—subsidies and loans as well as crop yields—were not evenly spread across society.

†The most potent criticism of genetically modified food, or genetic modification (GM) in general, it seems to me and many others, is that it is a prime candidate for the "law of unintended consequences" (see chapter 39). Critics of GM liken it to Pandora's box. It is easier to open than to close; any mistakes that occur when creating GM species cannot be undone once the modified species are released into the world.

‡The quote is from Folger (2014). For more on the original green revolution, and on the second version that is just getting under way, see Dowie (2001), the *Economist* (2014a), Hargrove and Coffman (2006), McNeil (2010), Snyder (2015), and Tierney (2008). See also the technical review of GM crops by Basu et al. (2010).

tion is environmental degradation (pesticide and fertilizer runoff, genetically modified organisms in the wild, and fossil fuel burning). On the other hand, returning to environmental sustainability may mean insufficient food for two billion of the people who will be born over the next few decades. Pick your poison—or rather, pick theirs.

 # Fat Americans

Americans are fat. Not every United States citizen, of course, but Joe Average. He is fatter than Fred Bloggs, his British equivalent,* or monsieur Jean Dupont from across the English Channel. He is fatter than Juan Pérez from Latin America or Mario Rossi from Italy. He is fatter than—I kid you not, this is the German equivalent of Joe Average—Otto Normalverbraucher. Or Vasya Pupkin or Zhang San. Americans are fat in the narrow sense of obesity ("narrow" is perhaps an inappropriate descriptor), but also in the broader sense of conspicuous consumption: more calories of food eaten, more watts of electrical power consumed, and more tons of carbon dioxide emitted.

First, the numbers for obesity. Over a third—35%—of Americans are obese today, up from 13% in 1962. Joe Average has put on 17 pounds since the 1970s, and Joanne Average is heavier by 19 pounds. There are three times as many obese American adolescents as there were back then. This growth in girth correlates with increased calorie intake: Americans eat more than they did 40 years ago, and they have always eaten more than people from other comparable countries (developed western democracies with roughly the same standard of living). Another major reason for increasing obesity is reduced exercise, especially over the last two decades. Today we don't even get up from our couch to switch TV channels.

American obesity varies widely with ethnicity, socioeconomic group, and state. The increasing numbers are leading to widened revolving doors, wider ferry seats, seat belt extenders on airplanes, ambulances retrofitted with winches, and wider coffins. Obesity is beginning to af-

*The British must also be getting fatter, if the craze for deep-fried Mars Bars (which I recall from Scotland in the late 1990s) is anything to go by.

fect the United States Army: in five years time, the number of obese young Americans will be so high that the military will be unable to recruit enough qualified soldiers. Obesity "is becoming a national security issue," according to Major General Allen Batschelet (Costello, 2015).

Average electrical power consumption in the United States is 1,683 watts per person. The average within the European Union is 688 watts. In China it is 458 watts and in India, 90 watts. Clearly, this disparity is correlated with living standards and economic development (so how much power will China and India be consuming a generation hence?). It also correlates with latitude: power consumption is almost as high as the US level in Sweden and Finland, and is higher in Canada. But Germany, Britain, and other countries with comparable living standards and climates consume significantly less power per capita than the United States. Humanity consumes about 20 terawatts; this figure will increase as population increases and as nations develop. Where will the power come from? If power supply becomes tight, will we be able to reduce our average consumption without compromising living standards?

Joe Average emits 17.0 tons of CO_2 into the atmosphere every year (not biologically—I mean due to his lifestyle. But you knew that.). Otto Normalverbraucher emits 8.9 tons; Fred Bloggs 7.1 tons. The numbers vary a lot from country to country and depend on the level of heavy industry and geographical size, but however you cut it, Americans are heavy emitters.[*]

The bloated figures—the numbers as well as the waistlines—for American consumption are good news in the sense that there is clearly a lot of slack in the system, if not in the belts. There is some room for Joe Average to reduce calorie intake, reduce power consumption, and reduce carbon dioxide emissions without adversely affecting his life-

---

[*]Obesity figures are from Costello (2015), Newman (2004), Kolbert (2009), and Snowdon (2014). See also the CDC website at www.cdc.gov/obesity/data.html. Electrical power consumption data is from Denny (2013) and the CIA *World Factbook* (http://theworldfactbook.info/index.html). Carbon dioxide emissions data are from the 2016 World Bank web article "CO2 emissions (metric tons per capita)." http://data.worldbank.org/indicator/EN.ATM.CO2E.PC?view=map.

style too much. He will likely have to make these reductions as he heads toward the Anthropocene.

• • •

We have encountered our first monster—not overweight Joe Average, since he constitutes just 5% of the human species—but our first planetary technological monster: hi-tech agriculture. (Four such monsters will grace these pages.) But is it the monster of *The Terminator* or of *Terminator 2*: sent to destroy us, or sent to save us? The emotive label "monster" is from French philosopher Bruno Latour, who likened it to another fictional creature—Frankenstein's monster. We will later see how these monstrous products of science and technology will define the Anthropocene.

Now some basic Earth science: the physics of climate and of climate change.

 Climatology 101

"Good morning to you on this first day of September, 2060. Welcome to all 225,000 students who signed up for this online course at U of Phoenix! I am Professor Albedo and I will be ably assisted by 1,000 overworked postdocs on short-term contracts who will manage tutorials and mark exam papers! To business: this short course introduces you to the fundamentals of climate science. We start, as all climatology courses must start, with the heat of the sun."

"Hack, hack! Hack, hack, hack! Apologies—I seem to have developed a nasty cough. All of the energy that drives our dynamic atmosphere and oceans—that moves all the fluids surrounding the surface of Earth—comes from the sun in the form of electromagnetic radiation: visible light, infrared (that is, heat), UV light, etc. The basic laws of thermodynamics tell us the *power spectrum* of this radiation—how much power is emitted at different frequencies.* This solar power bathes

---

*Professor Albedo is referring here to *black-body radiation*. With a few basic assumptions, physicists in the nineteenth century derived the physical law that shows how the temperature of a star depends on the power it emits (the *Stefan-Boltzmann law*). The power is distributed across radiation frequencies with a characteristic

the Earth and other planets; from the distance between the Earth and the sun and from the size of the Earth, we can work out how much power our planet absorbs from the sun. This power is pretty much constant from one millennium to the next, because the power output from the sun is fairly constant over that timescale and because the Earth's orbit is close to a circle, with constant radius."

"If we ignore global warming for a moment—we will see where it comes from in another lecture—then the planets emit at microwave frequencies all the power they receive from the sun. If they did not, they would have kept on heating up over the eons. In fact, they reach a constant *equilibrium temperature*, different for each planet. Physicists can calculate this temperature quite easily: the calculation is accurate for planets that do not have an atmosphere—atmospheres complicate the physics, as we will soon see. If the Earth did not have an atmosphere, our equilibrium temperature would be 254 degrees absolute— that's –19°C. Very cold. In fact, our average surface temperature is more like 288 degrees absolute—that's +15°C. Why the extra 34°C? Because of the *greenhouse effect* caused by our atmosphere. You will learn about the greenhouse effect in your tutorials." (See chapter 15.)

"In earlier geological eras, the chemical constituents of our atmosphere changed due to natural processes, and the differing composition of the atmosphere over the ages led to differing degrees of heating due to the greenhouse effect. So the Earth's surface temperature has changed from epoch to epoch: sometimes it is cold enough to plunge the Earth into an ice age, and sometimes it is warm enough to grow forests in Antarctica. This is all natural. But last year, 2059, geologists declared that we had entered a new epoch, the Anthropocene, defined by human actions altering the Earth's surface, and in particular, altering its atmosphere. Our industries have generated such a large amount of *greenhouse gases* (GHGs) that we are causing the greenhouse effect to ramp up: instead of 34°C of natural greenhouse warming,

---

spectrum. For the sun, the spectrum is centered around visible light frequencies. The Earth and other planets are also "black bodies" emitting radiation with the same type of spectral distribution, but being cooler than the sun, their spectra are centered around much lower (microwave) frequencies. Sometimes we talk about wavelength rather than frequency of electromagnetic radiation. These two properties of a wave are related, as follows: frequency multiplied by wavelength equals the speed of light, which is constant.

we are now getting 36°C, and it will soon increase further—the temperature rise due to human activity is going to exceed 2°C just because of the amount of GHGs we put into the air a hundred years ago, two hundred years ago—let alone what we are putting in now." (See chapters 15, 16, and 17.)

"We climatologists are getting good at modeling the physics of our climate. We can predict how the heat from the sun warms the Earth and the oceans, how some of it is reflected back to space by clouds. The warmed air and water circulate, like hot coffee in a mug, and so the heat is redistributed around the globe in complicated ways. There are different timescales for these movements: ocean circulations take from decades to millennia; atmospheric circulations are quicker. The existence of life on Earth complicates climate matters. To a geologist, trees are machines that fix atmospheric carbon: take it out of the air and stuff it into the ground. To an atmospheric physicist, trees are very efficient machines that take groundwater and turn it into atmospheric water vapor. Water vapor and carbon dioxide are both GHGs—so you see how climatology depends on biology."

"The accurate models that we climatologists construct are necessarily very complicated, because the underlying science is so complex, with many interdependent variables describing physical, geological, and biological processes. These computer models are called GCMs; the letters were originally an acronym for General Circulation Model, but nowadays the models are so all-encompassing that we say the letters stand for Global Climate Model. These GCMs predict that a number of feedback processes are going on in the atmosphere." (See chapter 17.) "For example, warming the atmosphere leads to melting of polar icecaps, which reduces the amount of incoming sunlight that is reflected back out to space (ice is very reflective), so more sunlight is absorbed, causing the atmosphere to warm further, thus increasing the melting of icecaps, and so on—a vicious cycle."

"When we heat up the atmosphere, the extra energy causes the dynamics to go faster. If the Earth warms a little then we can confidently predict a little change. If the Earth heats a lot then we are not so sure what will happen, due to chaotic effects and the nonlinear feedback mechanisms. If the Earth warms by 2°C or less from its preindustrial level, our GCM models tell us that the Earth will settle into a stable state, though the local weather will consist of more extreme events than we used to get—more droughts, more floods, more heat waves. If, however, the Earth heats up beyond a critical temperature called the *tipping point*, which is thought to be roughly 292° absolute temperature, that is, 19°C, or, 4°C above the preindustrial average global temperature, feedback mechanisms and chaotic effects kick in, and the atmospheric and oceanic dynamics may move into an unstable and essentially unpredictable phase. The resulting climate may even be incompatible with life, if we get the sort of runaway greenhouse heating such as happened on Venus."

"One last comment on GCMs for this lecture. Hack, hack, hack, hack, hack!—damn this cough. I need a glass of water. I said that our computer models are getting more and more accurate as physicists understand the underlying processes better and better, and as computing power increases. Climate modelers test their GCMs by *hind-casting*, which consists of "predicting" past climate changes. So they can start a simulation in the year 2000, say, and predict the climate of today—2060. They compare predictions with what is actually observed today, and get a pretty good match. This shows that the models work (and so we can be confident that the GCMs will also work when we apply them to the future climate). Of course, the climatologists had to put in by hand the huge climate effects that arose from the massive and widespread volcanic eruptions that occurred in the 2050s—you remember when Indonesia was wiped out and all that gunk ejected by the volcanoes caused the sky to go dark for three years. Oh, and they had to put in by hand the varying anthropogenic GHG emissions, which depend on industrial output and economics—recall the severe stagflation of the 2040s. Nobody could predict those eruptions or those economic cycles, and so they totally threw off the GCM predictions— but aside from that they were very good. Once you allow for the unpredictable stuff, once you feed it into the models by hand, then the GCM model predictions are pretty good."*

*For more details on blackbody radiation, the greenhouse effect, and GCMs, see Denny (2017) chapters 1 and 10.

15 Greenhouse Effects

Every chemical compound, in particular every gaseous component of our atmosphere, absorbs electromagnetic radiation at a rate that depends on the frequency of the radiation. The *absorption spectrum* of a substance such as water vapor can be summarized in a graph that plots the amount of radiation absorbed as a function of radiation frequency; this curve is a unique signature of the substance. So, for example, the spectrum for carbon dioxide differs from that for oxygen, or ammonia. A greenhouse gas is one that is found in the atmosphere and absorbs microwave and infrared frequencies (collectively known as *long-wave radiation*) much more strongly than it absorbs visible frequencies (*short-wave radiation*). The main GHGs in our atmosphere are water vapor, CO_2, methane, ozone, and CFCs (chlorofluorocarbons—including *freon*—organic compounds which were once used widely as refrigerants—see chapter **21**).

The significance of GHGs is that, because of their absorption spectra, radiation from the sun passes through the atmosphere on its way down to the surface without being absorbed much (because solar radiation is short wave). On the other hand, the long-wave radiation that is emitted by the Earth in response to this incoming solar power (see chapter **14**) *is* absorbed by the atmosphere. Thus the GHGs act like a valve for electromagnetic radiation: incoming short-wave solar radiation is passed, whereas outgoing long-wave black-body radiation is stopped. The net result is that the Earth and its atmosphere absorb more energy than they emit, and so they warm up.

These gases are given the label "greenhouse" because there is a fairly good analogy between their action and the action of glass in a greenhouse. Glass also passes visible light without much absorption (in other words, it is transparent) but absorbs long-wave radiation. Thus, heat is trapped in a greenhouse, and so the air inside becomes warmer than the air outside. This air is trapped by the glass, so the heat is not dissipated by convection. The analogy with a greenhouse is a good one, in that both employ the valve mechanism (GHGs/glass) to trap radiation. This analogy can be stretched to include the runaway greenhouse effect, a feedback mechanism whereby greenhouse gases become so prevalent in the atmosphere that the surface is heated up enormously, as on Venus. The Venusian greenhouse has many layers of glass (see, e.g., Denny, 2017). The greenhouse analogy breaks down when we consider convection, which is prevented by greenhouse glass but is an important driver of meteorological changes in our atmosphere.

Of the different GHGs, water vapor and CO_2 are by far the most common, so they make a more important contribution to global warming than do the other GHGs, even though some of these others are, molecule for molecule, more effective absorbers of long-wave radiation. Carbon dioxide matters more than water vapor, because the water vapor levels in our atmosphere are affected strongly by local temperature, whereas CO_2 levels are impacted by many other factors, including human industry. Why does this fact make CO_2 a more important GHG than water vapor? A warmer atmosphere can hold more water vapor than a cooler one, so the effect of adding CO_2 to the atmosphere is multiplied: more CO_2 means more atmospheric heating, because CO_2 absorbs long-wave radiation; this extra heating causes more water vapor to evaporate from the oceans' surfaces, which leads to additional heating, because water vapor is also a GHG. It has been calculated that the impact of CO_2 as a GHG is doubled or tripled as a result of this effect.

● ● ●

The levels of atmospheric CO_2 in earlier ages can be estimated accurately in several ways, most directly by analyzing bubbles of air that are trapped in Antarctic ice at known times in the geological past. Thus, ice cores—up to 3.2 km long—have yielded data about atmospheric CO_2 levels going back 800,000 years. During most of this period, levels varied between about 180 ppm (parts per million) and 270 ppm. The level that climatologists consider existed immediately prior to IR1, the first instance of human society industrializing, is 280 ppm. By 1960, the average level of atmospheric CO_2 had risen to 313 ppm, and in 2013, it passed the 400 ppm mark. The consequence of all this additional atmospheric GHG *so far* is an increase in the global average surface temperature of about 1°C. It takes a while for the effects of extra GHGs to manifest, because of the slow heartbeat of climate change—slow on the human timescale, that is. Carbon dioxide that is already in the atmosphere will continue to cause global warming for centuries to come—recall that it stays in the air for a long time. How much warming can we expect for a given mass of CO_2 added to the atmosphere? Global climate models show that if the preindustrial level of atmospheric CO_2 doubles (from 280 ppm to 560 ppm), then the Earth's surface will warm by between 1.5°C and 4.5°C. Note the wide variation: GCMs are good and getting better, but they are not perfect, because climate scientists do not have a perfect understanding of their complex subject. Needless to say, the undoubted facts of increasing atmospheric CO_2 and of global warming, coupled with quantitative uncertainty about the exact relationship between the two, is causing much debate that is, well, heated. More on the contentious issue of global warming due to atmospheric CO_2 in chapter **16**.*

16 Global Warning

"Welcome to the second U Phoenix lecture in our series on climatology. Once again, for those of you who missed the first lecture due to the ongoing power outages, I am Professor Albedo, and these lectures form the introductory course for the first term of this year, 2060, to the most pressing natural science subject of our day. Today we look at the early history of the IPCC. Most students will already have heard of the IPCC. For those who haven't, the Intergovernmental Panel on Climate Change is a United Nations body (working with the World Meteorological Organization) set up in 1988 to report on the scientific basis for climate change and the implications it will have for human life, and to make recommendations on how we can adapt to climate change and mitigate its effects. The panel is made up of thousands of climate experts from across the world. The IPCC does not itself conduct research into climate change, nor does it monitor the climate—all such research and data gathering are carried out by climatologists working in many climate research institutes around the globe. These institutes publish papers on observed climate change

---

*See Amos (2006) for a nontechnical account of ice core samples, Denny (2017) for a semitechnical survey of global warming due to GHG, and Held and Soden (2000) for a technical paper on the warming effects of atmospheric water vapor and CO_2.

(GHG levels, for example), paleoclimatology (such as ice-core sample measurements), climate theory, and prediction. The IPCC monitors this research, weighs it, and then draws conclusions about the state of the climate. Hack, hack! Hack, hack! I can't seem to shake off this damn cough."

"A significant part of the IPCC reporting is based on the predictions of computer models—the GCMs we talked about last lecture. As the experts' understanding of the climate increases, more and more aspects of climate modeling are included in GCMs, so these models become more accurate as they develop. Thus, from the very early days, the 1980s, GCMs included the thermal influence of oceans on the atmosphere by inputting measured ocean surface temperatures and then calculating the effects of these temperatures on the atmosphere. By the 2010s GCMs modeled the oceans in much more detail: the oceans were regarded as part of the system along with the atmosphere, and the effects of solar heating were calculated from first principles. The ocean circulations and the surface temperatures became model outputs, not inputs. A major challenge for GCMs in the 2010s was modeling clouds correctly. The action of clouds can contribute to global warming or global cooling, depending on cloud altitude and water content. Cloud formation and dissipation occur as a result of local conditions that were difficult for global-scale computer models back then to deal with—the scale of clouds was too fine, and the computer simulations were too coarse. By the 2030s the GCMs became detailed enough to include small-scale structure of the atmosphere/ocean system in such detail that cloud formation and dissipation is predicted by the models, quite accurately. Today our challenge is modeling volcanic eruptions and the human impacts on climate—we cannot yet predict these. All this is to say that the computer models are never perfect; they have known weaknesses but are improving all the time."

"There were dozens of GCMs around the world in the 2010s, just as there are today. They were developed and operated more or less independently, so they made more or less independent predictions. The weighted average of their predictions form a significant part of IPCC reporting. Reading these reports (the fifth was published in 2013, the twenty-eighth in 2059) we are struck by the attention to statistical detail; confidence levels are provided for each prediction, along with the change in these levels since the previous report (reflecting improved data and models). Climatology is an inherently statistical field, and each physical model of the climate makes predictions that are statistical, just as the meteorologists make statistical predictions about our weather (probability of precipitation tomorrow, for example). Our climate and weather are both chaotic systems with complex feedback mechanisms, so prediction accuracy decreases with prediction interval (projection into the future)."

• • •

"Let me summarize the IPCC conclusions about the state of the climate in 2013 (based on observations) and the direction in which it was heading back then (based on GCMs): *Warming of the climate system is unequivocal, and since the 1950s, many of the observed changes are unprecedented . . . The atmosphere and oceans have warmed, the amounts of snow and ice have diminished, sea level has risen, and the concentrations of greenhouse gases have increased.* They rated as "likely" (with "medium confidence") the assertion that the 30 years from 1980 had been the warmest in 1,400 years. This warming is "very likely" due to human activity, they said, and is "virtually certain" to continue at least to the end of the twenty-first century, they said. We now know this to be a slam dunk, from our current perspective in 2060. Let me continue summarizing the 2013 IPCC report. *Warming since the mid-twentieth century is "extremely likely" to have been caused by humans.* Damned right it was. *Heat waves and incidents of very heavy precipitation are both "likely" to be increasing in frequency and intensity, and this trend is "very likely" to continue.* Don't laugh; we have just sizzled through the hottest Phoenix summer on record and it killed a lot of ailing retirees; we have had water restrictions in the Phoenix area for over a decade. *Droughts are "more likely than not" to increase in duration and intensity. It is "virtually certain" that the upper layer of the oceans—the top 700 meters—has warmed since 1970, and "very likely" since 1870.* All true, we now know with near certainty. *Concentrations of three GHGs (carbon dioxide, methane, and nitrous oxide) are higher than they have been in the last 800,000 years, and the rate at which they are currently increasing is unprecedented in the last 22,000, with "very high confidence."* Ha! Of course, levels are even higher now."

"I will simply paraphrase the 2013 report summary, without comment. Remember that future warming trends depend on which scenario is adopted. For example, if we as a species had ceased burning all fossil fuels by 2020, the subsequent warming would have been less than it actually was, because in fact we continued to burn such fuels until mid-century. Most scenarios vary between extremes, from "cease burning fossil fuels immediately" to "business as usual." The IPCC concluded in 2013 that *global surface temperature increase by the end of the twenty-first century are "likely" to exceed 1.5°C in most scenarios* and will *continue beyond 2100.* The warming will *vary from region to region* and the rate of warming will not be uniform across the decades. This warming will penetrate from ocean surfaces to the depths ("high confidence") and will affect ocean circulation. *It is "very likely" that Arctic sea ice cover will continue to shrink and thin, and that Northern Hemisphere spring snow cover will continue to decrease. Sea levels will consequently rise."*

• • •

At this point I will part company with Professor Albedo. The rest of his course concentrates on the social discord and polarization that emerges during the 2050s, sown by the effects of increasing climate change around the world, which I can summarize in a few sentences. It seems that within the United States, the two main political parties fragment over the coming decades, so that by 2060, the Democratic Party has spawned the Kumbaya Mother Earth movement and the Vacuous Blowhard lobby, whereas the Republican Party, riven with internal disagreements over economic policy and immigration since the 2010s, has given rise to the Armed Wingnut brigade and the Glory Jesusland army. Let us leave these developments for the future, and return to the present day.

The variability in global warming is causing confusion and disagreement. We can expect climate warming to be variable, given the statistical nature of the beast, with this variability superimposed on a background rise for the last 200 years. All the more puzzling, then, that the data appear to show a pause in global warming over the last decade, covering at least the period 2004 to 2013. This pause was not predicted by most of the GCMs and is not fully understood.* Critics of

---

*The pause in global warming is now over, with record world average temperatures again recurring every month.

climate change, and in particular of anthropogenic climate change, pounce on this pause as evidence of flawed models or of overrating the human influence on global warming. The chief scientist who observed this pause says that it reflects our imperfect understanding of climate systems and is not due to flawed models. He expects an eventual 4°C rise in global temperature. The pause in global warming may have resulted from heat being diverted to the deep oceans; more data and further work are needed. The pause probably pushes off predicted changes a little so that, for example, the state predicted for the climate of 2050 will now likely not arise until 2065. At the time of writing, the pause has most definitely ended, and the 4°C prediction is being quoted increasingly in the media. (Recall from Professor Albedo's first lecture that a 4°C rise will take us near or beyond the tipping point for chaotic climate changes.) The current El Niño is warming the world: it is one of the strongest on record, though this was not foreseen— GCMs and numerical weather prediction models are not yet good enough to do so.

Despite this pause, some recent reports claim that the underlying global temperature rise is *faster* than we thought. It seems that those GCMs that most accurately simulate the observed changes in cloud cover (recall that clouds are a problem for current computer models of climate) also produce predictions for global warming that are at the high end of the range.

The situation is confused, naturally enough, by theoretical uncertainty about the physics, and this confusion is stoked by the high stakes—nothing less than the quality of human life. Warnings of global warming are treated with caution because most people have an imprecise understanding of statistics (see chapter **29**) and because they know the underlying science is not yet completely pinned down. The reporting of the IPCC and other climate research findings incite discord, to say the least, by trenchantly held views expressed in the media, ranging from "global warming does not exist" (Taylor, 2015) to "we're fucked" (Zolfagharifard, 2014). The fuzziness of the issues— despite increasingly clear evidence of warming trends—will tend to delay action, I claim, and thus make matters worse (see chapter **33**).*

---

*For reporting of extreme opinions regarding climate change, based on the evidence or otherwise, see, e.g., Booker (2008), Taylor (2015), and Zolfagharifard

⑰ 2 C or Not 2 C, That Is the Question

We saw in chapters **14** and **16** that climate models are now getting pretty good, though we will later (chapter **33**) be arguing that they can never be good enough. What do these models say about the consequences of current global warming (which will continue for a while yet, whatever we do about our contribution to the cause of it)?

There are a number of environmental consequences of a sudden (in geological terms) rise in temperature. Increased evaporation from the surfaces of oceans raises the atmospheric water vapor content—and water vapor is itself a greenhouse gas. A warmer atmosphere is boosted, like a kid on caffeine, to do what it does anyway, but faster and with more intensity. Rainfall increases and rainstorms are more intense; droughts are more widely spread and are drier. Storms are more common and more severe. Atmospheric circulation patterns are kicked up a notch. These consequences are not uniform around the globe and do not occur at the same rate, but they are inevitable. A key prediction that has emerged from models of climate dynamics—the GCMs—is that some of the predictions are linear and some are nonlinear. A linear feature changes in proportion to the global temperature change, so if global crop yields drop by 10% as a result of a 1°C rise in global surface temperature, they will drop 20% as a result of a 2°C rise. A nonlinear feature changes more wildly, often as a result of feedback. Thus if a global temperature increase is sufficient to cause Arctic ice to melt, then the exposed surfaces release greenhouse gases previously trapped under the ice, thereby increasing atmospheric greenhouse gases, thus increasing global warming further, and so melting more

(2014). For less extreme but still widely differing views, see, e.g., Carey (2012), Kirby (2013), Rose (2013), Nuccitelli (2015), and Pearlman (2014). The pause in global warming was first reported by Otto et al. (2013). See also Harrabin (2015), Snyder (2015), and Schwartz (2015). For more on the heating of deep sea waters, see Guemas et al. (2013), Gattuso et al. (2015) and Levitus et al. (2012). The latest IPCC report (IPCC, 2013) is available online. Much paleoclimatology proxy data is available at the NOAA website www.ncdc.noaa.gov/data-access/paleoclimatology -data/datasets.

Arctic ice, etc. Such a vicious cycle is called *positive feedback*—change leads to stronger change. Our complex climate provides examples of many interlinked phenomena exhibiting both positive feedback and its benign cousin *negative feedback*, through which change is resisted and stability reigns.

The GCMs seem to be telling us that there is a threshold, or *guardrail* as some researchers call it, at 2°C. That is, if we can keep the rise in global surface temperatures to within 2°C of the preindustrial level, we can limit the consequences—they are mostly linear. With more than a 2°C increase, however, some nonlinear phenomena kick in and may result in runaway effects and an unstable atmosphere. The precise nature and extent of such nonlinear behavior is difficult to predict accurately, especially into the far future (this is why weather is difficult—for all practical purposes, impossible—to predict a month ahead).

Much dispute accompanies the figure of 2°C. Some people think it is too low and others that it is too high; that we can achieve this limit or that it is impossible to do so; that any single measure such as this number is misleading as a realistic touchstone for climate change. It seems that the figure first arose in 1975, when Yale economist Bill Nordhaus, known for his contributions to our knowledge of the economics of climate change, intuited that a temperature rise of more than 2°C would take the climate outside the range it had been in for the last several hundred thousand years. In 1990 the Stockholm Environmental Institute suggested that a cap of 2°C above preindustrial global average temperature should be imposed by policymakers, a limit backed up by the European Council's environment ministers in 1996. In 2008, however, the United States cut reference to a 2°C limit from a draft G8 summit conclusion. Then in 2009, at the Copenhagen climate conference, leaders failed to agree on a deal to limit the rise to 2°C (114 nations recognized the scientific value of this limit, but the resulting accord was nonbinding). By the middle of 2015, a G7 summit reversed the earlier G8 result and backed the 2°C target, as did the Paris Agreement in December 2015. There has been much speculation in the news media about the value of this limit and whether we can, or even should, meet it.

Many experts think that we *can* take action to limit the long-term rise in global temperature to 2°C, but in order to achieve this desirable goal we must act very soon. If we delay deep cuts in emissions today, we will need deeper cuts tomorrow to keep to the limit, and we run the risk of a greater temporary rise in temperature before it falls back to the guardrail limit of 2°C. Thus, a delay of 20 years in taking action to cut emissions will require reductions of between 5 times and 7 times more to achieve the same long-term goal.

Others are saying that it will be difficult to meet the target whatever we do. Here is Professor David Victor of the University of California at San Diego, an expert on environmental regulations: "There is no scenario by which any accord that's realistic is going to get us to 2 degrees because the trajectory on emissions right now is way above 2 degrees" (Roach, 2014). The IPCC identified a carbon budget of 10^{12} tons; that is, humanity can burn a maximum of a trillion tons of carbon dioxide if it wishes to limit the global temperature rise to 2°C. We have, over the past 2 centuries, already burned 52% of this budget and are on course to burn through the rest in 30 years. We are seeing the effects today:

- Global average temperature is 1°C above historical levels
- Sea levels have risen almost twice as much from 1993 to 2010 as they did from 1901 to 1991
- The western United States today suffers seven times as many large forest fires as it did in the 1970s
- Heavy precipitation has increased
- The length and intensity of droughts have increased

The 2°C limit has political value—it is not a perfect yardstick but is a reasonable ballpark figure to aim for—though in the view of many doubters, it focuses effort on something that now cannot realistically be achieved. It may have been a sensible limit in 1990, but by dithering for a generation, the world has fewer options today. To stay within 2°C, we must achieve peak emissions by 2020 and leave three-quarters of known oil, gas, and coal reserves in the ground. This just isn't going to happen, not in developing countries (without binding protocols in place—see chapter **19**) and not in developed countries such as the United States, where the fossil fuel lobby is very powerful.*

---

*Thus Senator Mitch McConnell has accused former president Obama of waging a "war on coal" and has claimed that his climate agenda has caused irrevocable harm

Perhaps a 2°C rise is too much anyway. Recent reports say that when we get to a 2°C increase, the world will suffer bad climate impacts: many Pacific islands will become uninhabitable, and many millions of people will be displaced from coastal areas, due to higher sea levels; there will be more weather-related disasters such as heat waves, droughts, and floods.

●　●　●

Here are the known linear consequences of global temperature rise alluded to earlier. For each 1° Centigrade increase in global mean surface temperature there will be:

- a 5–10% change in precipitation across the world (whether it increases or decreases depends on location)
- a 3–10% increase in the intensity of precipitation events (such as rainstorms)
- a 5–10% change in river basin streamflow
- a 15% reduction in average sea ice coverage
- a 5–15% reduction of crop yields
- a 200–400% increase in the area of wildfire burns in the western United States

Other consequences are assured but are harder to quantify. For example, flooding will certainly escalate, but the damage it causes depends on factors other than sea level rise and precipitation rates, such as urbanization and infrastructure.

The nonlinear consequences of higher rises are worse, as you might expect. Thus, a 4°C gain is predicted to lead to a collapse of farming in sub-Saharan Africa and the extinction of many species of plants and animals, plus a third more of the world's population will face reduced groundwater resources by 2080 (compared with 1980). At the same

---

to the coal industry in Kentucky. He urges states not to comply with emission reduction pleas from the Obama administration and calls on world leaders to be skeptical of Obama's commitments. President Donald Trump has claimed that climate change is a hoax put about by either China or climate scientists. If such attitudes prevail in the corridors of power, it is hard to see how GHG emissions in the United States will be reduced over the coming decades.

time, there will be much more flooding. Note that predictions of the human consequences of high global warming depend on knowing what the population will be in the future and how it will be distributed around the world—see chapter **11**. Other predictions are very long term and so are subject to prediction errors. Thus, warming the deep oceans will release carbon that is currently trapped in deep-sea sediments, but this will take centuries, and its effects are hard to put a number on.

•   •   •

There is more carbon dioxide in the atmosphere today than at any time in the last 800,000 years, as we have seen (chapter **16**). Fifty-five percent of added CO_2 is promptly absorbed by plants, soil, and the oceans. The remainder stays in the air for a long time; more than half is still there a century later. The current level of atmospheric carbon dioxide (about 400 ppm) is compatible with a 2°C rise, but if the amount increases to 450–750 ppm, the rise will likely be closer to 3°C. At 1,000 ppm, the temperature surge will eventually level off at 5°C. The timescales of global temperature change are slow; the carbon we have dumped in the air over the last 200 years has not yet led to a 2°C step-up, but the wheels are turning, and a rise of at least that amount is inevitable. Whether a 2°C hike is achieved temporarily, is the new long-term average global temperature rise, or is a marker missed as we head to much warmer times in decades and centuries to come, depends on the levels of carbon dioxide we burn today and tomorrow.*

*See, e.g., Friedman (2015), Foran (2015), Goering (2015), Kriegler et al. (2013a,b), Mooney (2015a), Nijhuis (2014b), Roach (2014), and Sutter (2015). See also the special issue of *Philosophical Transactions of the Royal Society* (January, 2011). There are also many online articles and reports, such as the World Resources Institute website, which includes the *IPCC 5th Assessment Report* on the effects of emission reductions, from best case to worst case scenarios. Another online article worth reading is the National Research Council report *Climate Stabilization Targets: Emissions, Concentrations and Impacts over Decades to Millennia* (2011), http://nas-sites.org/americasclimatechoices/other-reports-on-climate-change/climate-stabilization-targets/. Increasingly, reports coming in suggest that we're heading beyond a 2°C escalation.

18 Tipping Points and Tipplers

As an experienced popular science writer, I find myself often called upon to conjure up an analogy. Deep understanding of a science concept, especially physics, usually requires mathematics, because math is the only language we have in common with Mother Nature. Most people don't speak math, however—it brings to their minds unloved lessons in high school, with long-forgotten rules of trigonometry and mysterious algebraic scratchings on chalkboards. So to explain science ideas, I and many other science writers resort to analogies. At best, analogies convey the essence of the problem being explained: saying that a sail or a bird wing operates like an airfoil is a good analogy—one that stands up to detailed scrutiny. At worst, however, analogies can be seriously misleading: saying that electrons orbit atoms like planets orbit the sun is a common but very bad analogy. The notion of an ecological or climatic tipping point is a complex natural phenomenon that calls for a good analogy, provided herein.

Climate modelers have a precise notion of what "tipping point" means and the forms that it can take; these forms can be easily expressed in a few words by writers of popular science, for example, "runaway global warming," but the precise meaning is harder to convey. "Positive feedback" or "vicious cycle" get us part way there; they cover the facts by throwing a large blanket over them (to employ an analogy) but are consequently too vague. I might attempt to be very precise by providing an actual example, say by calling the manmade environmental disaster that occurred on Easter Island an environmental tipping point. But this historical analogy is not a good one and misses the mark (see chapter 20).

Here is my analogy; I apologize in advance to readers of a sensitive disposition for its coarseness, but really, this is the best I can do. We—you and I—are drunks at a party with a cash bar, sprawled against a wall, guzzling bottles of beer. We are elegantly dressed but sitting in pools of vomit amid broken bottles. A perverse host has placed the beer behind a faux brick wall. We must remove a brick to obtain a bottle. We dispose of a brick each time we obtain a bottle, as follows. There is a fragile shelf on the wall above us; each time we get a bottle of beer, we throw a brick up onto the shelf. As we sit below, happily if somewhat noisily glugging beer, the shelf above us groans under the ever-increasing weight of bricks. I think you can see where I am going.

This is a good analogy, to my mind, for several reasons. The high-consumption societies of the world (see chapter 13) are represented by smartly dressed drunks at a party, happily careening toward an unpleasant future through stupidity and their own actions. When sober, they might realize the danger, but they're not and so they don't—they are having too good a time. Other guests at the party realize what's happening and are trying to moderate the overdrinkers' behavior, but it is difficult and unrewarding to give bad news to a drunk. If the partypoopers are teetotalers then they will be derided, and if they are imbibers they will be offered a drink or told that they are hypocrites. The supply of fossil fuels—by which I mean beer, of course—may or may not be limited, but for our purposes this detail is unimportant if the shelf gives way before the beer runs out; there are terrible consequences to our overconsumption. Should we stop drinking (give up fossil fuels) or stop throwing bricks (eliminate harmful emissions)? A *tipping point* is a point of no return, an irreversible consequence of recent events. Here, clearly I hope, the shelf collapsing is analogous to a disastrous environmental tipping point.

It is amusing to stretch this analogy further. Suppose you (one of the drunks, let's face it) retain a small part of your ability to think clearly and realize the danger that your actions and those of your fellow tipplers are causing. You see that the shelf is loaded to its breaking point—one more brick will cause it to fail. You see me in the act of hurling another brick. You reach out and halt my moving arm, with the brick in it. Is this where humanity is at, today? Consumers taking just enough action, just in time, to prevent an environmental tipping point? Or maybe the last brick has left my grasp in the instant before you stay my arm, and is inexorably on its way to the shelf? After all, there is a time lag in climate physics, and perhaps it is too late already to avoid a tipping point, even if we now realize our past mistakes and amend our profligate ways.

Others at the party may get drunk without throwing bricks* or may not be able to afford the bar bill. They may not be under the shelf and so may not suffer the full consequences of its collapse. However, the party just won't be the same afterward. The room will be a mess, and the party will be spoiled because there is no other room available (i.e., no other planet for us to live on). Enough—you get the picture.

Climate Change Protocols

Nations in the process of industrializing want to be as rich as the already-industrialized countries of the western world, naturally enough. Some developing nations suspect that the developed world is resisting this industrialization, that we want these countries to stay poor because we fear the competition or the increased clout that comes with a burgeoning economy. These fears may or may not be well founded, but what is certainly true is that much of the developed world is resisting the GHG emissions that have been an unavoidable product of industrialization since the early 1800s. This resistance arises from the increasingly prevalent viewpoint that our climate is changing in ways that will be bad for humankind and for much of the rest of the biosphere—changes that are being driven in general by human activity and in particular, by heavy industry.

The evidence for human driven climate change is now so overwhelming (see chapters **16** and **17**) that only the most narrow-minded vested interests choose to deny it. The problem now is what to do about it. Not only must we in the developed world reduce our own emissions and move to industrial practices that are compatible with climate stability, but we have to convince the BRICS nations and other industri-

---

*Norway is a consumer society fueled mostly by clean hydroelectric power—I suppose that Norway may therefore be considered as one of the party drunks (high-consumption countries) who is not throwing bricks, in this analogy. Warming to this theme, I note that Mr. USA is the loudest drunk at the party, cluelessly throwing his weight around and buying beers for others; Mr. Britain has more or less passed out—he started drinking earlier than everyone else; Ms. China and Ms. India are the heaviest drinkers, guzzling like there is no tomorrow . . .

alizing countries that they, too, must cut down or eliminate fossil fuel burning. Of course, the leaders of these nations point to us with the obvious contradiction of our standpoint: "you in the developed world created the climate change of today by burning coal for the last 200 years; what right do you have to deny us economic development by the same route?" Fair point. True, we did not know what we were doing back in the day—people in nineteenth-century mills certainly knew about locally blighted landscapes (recall chapter 8) and unbreathable air, but they had no idea that they were changing the global climate. However, the fact is that much of the carbon dioxide in the atmosphere was put there a century ago; it lingers. Industrialization today, powered by fossil fuels, will belch out more CO_2 that will remain in the air for the next century, giving rise to more rapid and extreme climate change. So the nations of the world gather to see if and how they can reach an agreement by which developing nations can industrialize without burning fossil fuels to the extent we did when we industrialized, and to see how developed nations can drastically cut their emissions without harming their economies.

This is a big problem, and one that would be tough to solve even if nations and people were altruistic. But neither people nor nations are altruistic in general. Many developing countries are not pluralistic, and their leaders care little for the effects of industrial pollution on their own people, let alone the rest of the world—they want to industrialize in the fastest and least expensive way possible, and that means burning coal. Many western countries are democracies with leaders who think no further ahead than the next election; such timescales are much shorter than some of the climate change timescales—why not just push the bad effects of pollution into the next administration—kick the can down the road?*

• • •

*I recall a debate in the mid-1980s about polluting smokestacks in Britain. The solution was to increase the height of the smokestacks so the prevailing winds would carry the pollution out to sea. Subsequently this pollution was claimed as the cause of acid rain in countries downwind, principally Norway and Sweden. This case illustrates the length of time that pollution issues have been in the international spotlight—at least three decades—and shows an obvious selfishness or shortsightedness, even in countries where pollution is a major issue. See Caulfield (1984).

Perhaps for these reasons, or perhaps because the emission-control nut really is a hard one to crack, we have failed to reach an effective collective agreement since the urgency of the climate change situation became evident to world leaders in the 1990s. The first international treaty to reduce greenhouse gas emissions was the Kyoto Protocol of 1997. One hundred and ninety-two nations there agreed that global warming was due in large part to anthropogenic carbon dioxide. Unfortunately, the agreement has had little effect: the United States signed the Kyoto Protocol but did not ratify it; China signed and ratified it, but then exempted itself from it; Canada signed it, ratified it, and started to implement it, but a later, more business-oriented government withdrew from it altogether in 2012. To date there are 83 signatories, of whom 37 have committed to binding emission control targets.

A 2009 climate summit in Copenhagen ended disappointingly with a U.S.-China disagreement. These nations are the two biggest polluters, accounting for 40% of the world's CO_2 emissions, so their participation in any emission control protocol is vital. There was better news in 2013, when they agreed to phase out production and use of chlorofluorocarbons (greenhouse gases used in refrigerators). Following emission control negotiations in Lima, Peru, in 2014 China, India, and the United States said they would not ratify any treaty that committed them legally to reducing their carbon dioxide emissions. The European Union argued for mandatory emissions limits for all countries, and has agreed to binding cuts of 40% (compared to 1990 levels) by 2030. The United States wants to retain the freedom to adjust the scale and pace of emission control. Yet former president Obama's administration has reached significant deals with both China and India. The United States has pledged to cut emissions between 26% and 28% below 2005 levels by 2025, while China has promised that its emissions will peak by 2030 and has agreed to increase the fraction of nonfossil fuels to 20% by 2030. (Latest Department of Energy figures show China's coal consumption down 2.9% from the 3.8 billion tons burned in 2014—the most of any nation—though this might be due to a slowdown in the Chinese economy or to unreliable statistics gathered from provincial industries.) Meanwhile, US-India climate change talks have resulted in a heavy American investment into Indian solar and wind energy projects, but resulted in no statement of binding emission control targets or of a peaking year for Indian emissions.

The 2015 climate summit in Paris has been hailed, in its immediate aftermath, as a success. A deal has been reached that aims to cap global warming *below* the 2°C guardrail; this deal has been signed by all participating countries (nearly 200), but many of the commitments are not legally binding. There is a strong review mechanism to assess progress, and a clearly stated method for attaining the limited temperature increase (limiting anthropogenic GHG emissions). There is money available for helping poor countries to adopt climate mitigation measures.

Some analyses of the current (2017) level of commitments to emission control say that, even if they are fully implemented on time, they are not enough to hold the level of global warming to 2° Centigrade (see chapter 17) but will instead result in a rise in global temperature to 3.5°C or 4°C above that of the preindustrial world. This, climate models tell us, is very bad news.

The level of complexity of the emission control negotiations can be appreciated from a suggestion made by India in 2014, that emission controls should be set by consumption levels as well as production levels. That is, consumer nations are contributing to greenhouse gas emissions by buying products that required burning CO_2 to make them. By this measure, for example, the United Kingdom has reduced its direct production of greenhouse gases by 19% between 1990 and 2008, but its consumption has led to a rise in emissions in other parts of the world by 20%. This notion outsources the emission control problem, adding to the existing debate about how the cost of emission control should be paid. If industrializing countries are to be denied cheap coal, then perhaps industrialized countries—who got that way by burning coal themselves—should help pay for the cost of weaning the emerging nations off fossil fuels. Maybe the Paris agreement is a step in this direction. The questions are: who should pay and how much? Who should benefit and how much? Which nonfossil technologies should replace fossil fuel extractions in which countries? How should producers be compensated for leaving coal, oil, and gas in the ground?

The tangled issues have meant that little has been done in three decades. The Paris meeting produced meaningful commitments, but only time will tell if they will be implemented, and time is running out: kicking the can down the road will make the problem harder to fix, with deeper emissions cuts necessary for the same end results. The world is not a single nation with one person in absolute control and never has been, thank goodness. But with nobody in charge, a world without emissions control is like a NASA moonshot without Mission Control. We need a deal, and fast, but leaders are short sighted, and nations are looking out for themselves. (See chapters **30** and **40**.) It was ever thus, but now this fact of human nature is going to bite us.*

• • •

The numbers depend on whom you ask—they are disputed—but the trend of the evidence is increasingly pointing to a world that is warming to levels that will adversely affect the quality of human life. Perhaps, if we do nothing to counter the warming trend, global average surface temperature will attain levels that lead to significant reductions in our quality of life. Our best physics is put into detailed climate models, but the climate is so complex that we do not yet have a handle on all of it. Predictions about where past GHG emissions will take us in future decades are uncertain, in part because of this and in part because of future unpredictable influences such as volcanic eruptions and economics. And what about future emissions, during the Anthropocene—how do we predict their levels? We will see that we cannot.

First let us step back, and present some goodish news. The unfolding effects of climate change and of environmental degradation brought about by human actions have been likened to those that occurred on Easter Island over the last few hundred years. Does the resulting disaster for the people on that island tell us something about our own future? In fact the analogy is a false one, as we are about to see. Then, a good-news story about how timely and appropriate human action is effectively countering a serious environmental problem of our own making—the ozone hole.

20 Rapa Nui Not

For decades, academic research, popular science books, and Hollywood movies suggested that the geographically isolated Easter Island (a

---

*To find out more about emission control protocols and international efforts to reduce greenhouse gas emissions, see, e.g., *Boston Globe* (2014), Collyns (2014), Gallucci (2015), Gibb (2015), Lean (2014), and Yeo (2014).

63-square-mile speck in the South Pacific) had suffered environmental catastrophe as a consequence of the behavior of its human
inhabitants. These people, Polynesians who called their island *Rapa
Nui*, constructed mysterious statues—giant heads of stone they called
moai—facing out to sea. It was said that, to move these very heavy
blocks of stone to their coastal platforms from the inland quarry sites,
the people cut down trees and used the logs as rollers.* Hence, so the
old theory went, the land became deforested and erosion did its worst
and made the island uninhabitable. Easter Island was first occupied
by humans around 500 AD, and the population is reckoned to have
built up to a maximum of around 15,000, but due to environmental
degradation it fell to under 3,000 by the time the first Europeans arrived in 1722. Diseases brought by the Europeans plus slave raids from
Peru then finished off this once-thriving and sophisticated civilization.
The population was reduced to 111 by 1877.

Jared Diamond cites the Easter Island civilization as the "clearest
example of a society that destroyed itself by overexploiting its own resources." An old eco-activist website refers to the Easter Island story
as a "grim warning to the world." So perhaps the fate of this island—
it is still more or less denuded of trees and is not capable of supporting
even a small fraction of the population it once did—can be regarded as
a forewarning of what will happen on a larger scale. If we continue to
overexploit our natural resources, our fate will be that of the Rapa Nui
culture: population crash and social disruption—people fighting for the
remaining dwindling resources as the last trees fall and fertile land is
washed away into the sea.

However, recent research suggests that it didn't happen like that.
Current thinking is that the introduced Polynesian rat had more to
do with deforestation than did human activity. There were population
shifts within the island, as people moved from the interior and the
leeward side (probably due to changes in weather patterns over centuries) but little evidence to suggest a population crash before Europeans arrived. So we can blame rats, disease, and Peruvians for the
disaster that befell the Easter Island people. The environmental deg-

---

*Some people, sadly, learn their history from Hollywood movies—in which case the
1994 movie *Rapa-Nui* will have perpetuated in their minds the wrongheaded notion
of chopping down the last Easter Island tree for rollers.

radation of their island was probably not due to overexploitation. It seems that this isolated ecosystem was unusually fragile, so its collapse was pretty much inevitable as soon as it began to be stressed by humans (or rats).

It was a bit of a stretch anyway—a bad analogy—to extrapolate from Easter Island to the whole world, from chopping down trees for cultural reasons to global ecosystem destruction due to human industry. It may or may not be the case that we are ruining the surface of our planet as a habitat for humanity, but it is a false analogy to point to Easter Island as a precursor, as some have done. I am not saying that we shouldn't be warning governments and industries about the consequences of their actions, but I am saying that we shouldn't claim that the consequences of these actions will be the kind of destruction that, sadly, engulfed the people of Rapa Nui.*

21 Ozone Whole

Ozone is a gas, each molecule of which is formed from three oxygen atoms (O_3) rather than the two atoms that form oxygen gas (O_2). Ozone is less stable than oxygen, though the level of ozone in the stratosphere is—or was, until the human era—more or less constant. One of the characteristics of ozone is that it absorbs much of the incoming ultraviolet light from the sun. UV radiation is harmful for humans and other life, so the existence of a stratospheric layer of ozone (some 15 to 30 km above the surface) is useful. By absorbing this radiation, ozone molecules heat up the stratosphere; indeed, this layer of the atmosphere is defined by an increase of temperature with altitude—opposite to the temperature profile in the atmospheric layers above and below. (Thus in the troposphere—the lowest layer of the atmosphere, between the surface and the stratosphere—temperature typically falls by 6.5°C for every kilometer increase in altitude.)

---

*A lot has been written about the demise of the Easter Island population. See, e.g., Bressan (2011), Diamond (2005), Krulwich (2013), Pakandam (2009), Ponting, and Stevenson et al. (2015).

In the 1970s, members of the British Antarctic Survey noticed with alarm that the stratospheric ozone near the South Pole was thinning. This depletion of ozone was monitored and was found to be seasonal, being particularly noticeable in spring and summer and disappearing in the winter months. Over the years, researchers around the world built up data on the changing ozone levels. Stratospheric ozone was decreasing more and more each spring in Antarctic regions, and was thinning (but less rapidly) at other latitudes. Soon the levels near the South Pole were so thin that there was effectively a hole in the protective shroud of ozone at these latitudes, and the size of this hole was increasing year by year.

Research in the late 1970s established that humans were the cause of this ozone depletion. Chlorofluorocarbons (CFCs) are manmade gases that were widely used in refrigerators, air conditioners, and aerosol sprays. Once released into the air, the CFC molecules found their way up to the stratosphere, where they were broken down by UV light, releasing chlorine gas, which reacted with ozone, reducing it to ordinary oxygen. Publication of this research led to popular pressure on the large chemical companies that manufactured CFCs to get rid of them—to find safe alternatives. This ozone research was challenged by the chemical companies, who called it "speculative science" and who claimed that regulating the use of CFCs would lead to the loss of many jobs.

The ozone hole continued to deepen and spread throughout the stratosphere, as the arguments over CFC use deepened and spread over the surface below. Long story short, the Montreal Protocol of 1987 phased out CFC production and use. The treaty ban was signed by every member country of the United Nations—the first and only time that unanimity has been achieved. The prompt and immediate dangers of increased UV radiation—which causes skin cancers and cataracts in humans—perhaps spurred the legislation, along with the fact that replacements for CFCs were developed inexpensively.

The ozone hole continued to grow for the first two decades after CFCs were phased out and reached a maximum size of 30 million square kilometers in 2006. Since then, the steady annual growth of the hole has stopped and it is now shrinking, according to the latest estimates by the World Meteorological Organization; experts expect that most of the globe will return to 1980 levels of ozone by 2050. The

Antarctic hole will take a little longer to fill, but it is expected to be back to 1980 levels by 2075.

There is little doubt in the scientific community that manmade CFCs were responsible for most of the atmospheric ozone depletions, though the details are complicated (depletion rates vary with altitude as well as latitude, for example). There is no doubt that if CFCs had not been phased out, the ozone depletion would have spread around the planet, and the protective shroud would have disappeared. A bonus is that CFCs are GHGs; the Montreal Protocol has led to the reduction of five times as much GHG from the atmosphere as did Kyoto. Mario Molina, who received a Nobel prize for his ozone research, said the elimination of CFCs from our atmosphere was "a victory for diplomacy and for science and for the fact that we were able to work together."*

The resonance here may already have occurred to you. The bleating of industry, objecting to making changes that would benefit the environment; the arguments and disputed science; the hope that a similar victory might be achieved over global warming. Yet there were two factors that made the ozone hole problem an easier one for humans to collectively solve than the global warming problem. First, the risks to humans were more immediate: skin cancers and cataracts threaten us, not those who come after us, and in ways more readily envisioned than a warmer atmosphere. Second, the cost to fix the problem turned out to be small and took only a few years to develop. Even so, the optimists among you may draw some comfort from the fact that the ozone hole problem showed that humans can cooperate to clean up at least the simpler environmental problems we cause.

• • •

So, is there room for cautious optimism regarding climate change? It is becoming increasingly clear to more and more people that we need to perform considerable remediation to fix an environmental mess that is almost entirely our own creation; we have seen that it is possible to reverse some bad practices

---

*The quote is from Sullivan (2014). See also Biello (2014), Handwerk (2010), and Yang et al. (2006). There is a detailed World Meteorological Organization preprint available online that includes much useful data: *Scientific Assessment of Ozone Depletion 2014.*

that led directly to climate change—closing the ozone hole we created. We will see later what the likely climate developments will be, over the next decades, if we don't do anything to counter global warming, and if we do. The trouble is, the very concept of an Anthropocene age implies by definition that humanity significantly influences the environment and thus the climate; therefore to make meaningful predictions about the climate a century hence, we need to know about human developments between now and then. Such developments are unpredictable and consequently, it seems to me, our bright prospects are dulled. But I am getting ahead of myself—I will let this thesis develop further before hitting you on the head with it.

It is time to survey our current and future energy sources. We have seen in chapter 13 that energy needs vary widely from region to region; what will these needs be in the generations to come? What is the best way forward—and is it the same as the most probable way forward? There are now, and will be in our future, three types of energy sources: the Good, the Bad, and the Ugly. Let us consider them in order.

22 The Good

Renewable sources of energy are considered to be good because we can employ them into the indefinite future, and because they are much less damaging to the environment and climate than are the bad energy sources. Here I will briefly survey three of the most promising renewables, one of which is already a mature and well-established industry—the other two are in their adolescence. All three have serious limitations, however, and while they will play a significant and beneficial role in supplying the future energy needs of humankind, theirs will not be a dominant role until the second half of the century.

• • •

The surface of the Earth is bathed each year in 500,000 cubic kilometers of rain—that's about 120,000 cubic miles. Rain is a renewable resource, and hydroelectric dams are the means by which we trap the rainwater and funnel its gravitational potential energy into useful electric power. Dams have been built for millennia for the purpose of irrigation, or to keep water out of low-lying regions, but dam projects for the primary purpose of generating electricity have been built only over the last century. Rain is free but dams are not—large dam projects can break a nation's budget, such is their cost. They can also solve a nation's power supply problem; Norway gets all its energy and power* from dams, and Canada gets over half; 3% of the world's power is from this source.

Hydropower is clean—its output is electricity and water. Its cost per kilowatt-hour (a few cents) is lower than that of fossil fuels. Sounds ideal, doesn't it? Well, yes it does, if you can afford to build and maintain the dams and the electrical infrastructure. (Dams are built amid mountains or gorges, often a long way from population centers, and so significant lengths of electric power lines are needed to transfer the power from dam sites to cities.) The main limitations of hydropower, since the technical problems of building large dams were solved a century ago, is cost.

And geology and geography. To build a dam requires the right topography, bedrock constitution, and rainfall. This combination of natural environmental prerequisites limits the development of hydropower to certain regions of the globe. In this sense, hydropower is as quixotically distributed as oil. Thus, while Canada generates 58% of its power from hydro, neighboring America gets only 7% from this benign source. Within the Unites States, different states benefit to widely varying degrees: 90% of the power generated in Washington is hydropower, whereas 0% of Mississippi power is hydro. Developed countries have pretty much maxed out their hydropower already; it will continue indefinitely, but it will not grow much in absolute terms (number of megawatts produced) and may fall as a percentage of power needs as populations increase. The main growth in hydroelectric power generation these days is in developing countries, because they can now afford to build the dams and infrastructure. Thus China boasts the largest hydroelectric dam in the world—the Three Gorges Dam, with a capacity of 22.5 gigawatts (GW)—and China is largely responsible for East Asia overtaking both North America and Europe in terms of installed hydropower capacity. Only Africa remains largely untapped,

---

Energy is the stored quantity; when it flows it is measured as *power*.

with perhaps as little as 10% of its potential hydroelectric power realized.*

If all the world was like Scandinavia, we could have Abba, Ikea, and as much clean hydroelectric power as we can use, but it's not. At most, our world will be able to generate a couple of terawatts of hydropower, and that is barely 10% of current global demand, let alone what we will need in the future. So hydropower is good, but will not be enough to solve future needs except for a few lucky countries.

•   •   •

The sun bathes the Earth in electromagnetic radiation—sunlight, infrared, ultraviolet—to the tune of 180,000 terawatts. Humanity currently consumes about 20 terawatts of electric power. Do the math, and it seems like a no-brainer that we should invest some effort learning how to tap into all that free power, given to us every second of every minute of every day, by that beneficent thermonuclear reactor up in the sky—our sun.

Of course, solar power technology is not so simple. Take into account the curvature and rotation of the Earth, and deduct 30% of the radiation that reflects off clouds back into space, and we find that the average power density reaching the surface amounts to 240 watts per square meter. Nature taps into this energy source: photosynthetic plants are able to absorb about 1% of it. An early (1970s) solar panel was similarly efficient, but nowadays they are much better, absorbing about a sixth of the solar power that impinges upon their surface. In the next few decades, as technology improves further, this efficiency may increase to two-fifths, or 40%.

Photovoltaic (PV) solar power plants consist of vast arrays of these solar panels. There is another technology: *solar thermal electric* or STE, which concentrates the sun's rays via mirrors to cause a conductor to heat up. The stored heat is then used to generate electricity. STE is in its infancy compared to PV; it will likely come into its own around

---

*For more details on various aspects of hydropower, see Denny (2010) chapter 6, Denny (2013) chapter 6, Edwards (2003), Jackson (2005), Khagram (2004), and Ray (2010). See also any of the International Energy Agency (IEA) websites on hydropower.

the middle of the present century. Following the oil crisis of the 1970s, and especially following the nuclear accidents at Three Mile Island, Chernobyl, and more recently at Fukushima Daiichi, much effort and money have been invested in solar power technology, especially PV. This money took the form of feed-in tariffs and tax breaks, with the aim of reducing the cost of electricity generated by solar power to the level of fossil fuel electricity. Germany has been prominent in sponsoring solar power research in this way, following a decision in 2011 to abandon its nuclear power program, after noting the destruction caused by the Fukushima-Daiichi disaster. Many other countries also invested in PV technology, resulting in a solar generating capacity today of 305 GW—about 1% of electric power generated by humans.

The rise in PV capacity in recent years has been phenomenal. A third of all our PV capacity was installed between 2010 and 2012. It is the fastest-growing source of electricity in the United States and China, having more than doubled over the two-year period 2015–2017 as costs fell. The rest of the world, too, is going solar: Germany currently generates 40 GW, Italy 20 GW, China 77 GW, and Japan and the United States each 43 GW. Saudi Arabia is investing heavily in solar power—it makes sense for the desert nation when its oil runs out. This heady rise has led to some optimistic predictions of future solar power capacity: the International Energy Agency expects that, if investment and technological advances continue, solar power will constitute 27% of global electricity generating capacity by 2050 (16% PV and 11% STE). Other predictions are off the wall: solar power will drive the world in twenty years; it will be unlimited and free.

Unfortunately, the exponential growth of recent years may stop dead, at least for a while, because the funding that has caused it is drying up. Germany drastically reduced its feed-in tariff in 2012, because it was simply too expensive; much of the rest of Europe followed a couple of years later, and R&D money in the United States is also drying up. The cost of a solar panel has indeed plummeted as a result of advances in manufacturing technology, and the cost per kilowatt of electrical power today is comparable with that of nonrenewable sources, so perhaps solar power (the PV type, anyway) can now fly on its own. Time will tell.

Also, it is worth noting that PV solar power is not quite an environmental free ride: true, it generates zero carbon dioxide, but there are plenty of toxic chemicals used in the manufacturing of panels; disposing of old panels—there will be hundreds of millions of them, if this technology achieves its potential—will need to be done with care. In addition, the land area occupied by solar power plants may be problematic for some countries. The distribution of solar power varies with latitude—near the equator, the incoming solar power density may be four times the 240 watts per square meter average, whereas northern countries under cloudy skies may get only a third of the average.* And of course, solar power shuts off at night; the problems of integrating such a variable power supply into national electricity grids is raised in chapter 26. Sunlight may provide a significant source of electricity in the future of some countries, but not for the world as a whole. I see no reason to change the prediction I made a few years ago: that solar power will account for around 10% of global electrical power by midcentury, unless the price of solar panels continues to drop, to the extent that solar power becomes feasible for installation on the rooftops of developing nations; in that case, perhaps 20% of world electricity generation may occur in large-scale utility and small-scale household sunlight farms.†

• • •

While solar power is seen by many as the brightest hope for future renewable energy, we can also take heart in the way that the winds of change are favoring another green technology. Wind power is very hi-tech, and always has been. In medieval Europe, windmills were technological marvels and constituted the cutting edge of power generation, in an age when power was needed only for grinding corn or

*So why has cloudy Germany invested so heavily in solar power? Perhaps the Germans are playing a long game—by the end of the century the German climate may be Mediterranean, due to global warming. See chapter 36.

†The IEA claims for the future of solar power are outlined in Van der Hoeven (2014). See also IEA (2015). For other reasonable and unreasonable recent accounts of solar power, see Cusick (2015), Denny (2013) chapter 8, Harrington (2015)—a good example of bad solar power math—Miller (2015), Thomson (2014), and Raj (2014).

pumping water. These impressive machines remain today as architectural curiosities amid northern European plains covered with much larger wind turbines, generating much more power for people with much greater power requirements. These turbines are the result of an enormous theoretical and practical engineering effort to capture energy from the wind, channeling the kinetic energy of air movement into electrical power.

For wind power generation, size matters. The efficiency, energy density, and cost per kilowatt of power generated all benefit from increased size—by every measure it makes sense to go big. In 1999, the average length of a rotor blade was 24 m, and the blades were attached to a hub that was 55 m above the ground. By 2012, the towers that supported the rotor had grown so that the hub was 93 m above the ground and the rotor lengths averaged 43 m. This increase in scale means that an average wind turbine has increased in power by a factor of 4, just by scaling up, not including any improvements in efficiency or design that occurred during these 13 years.

The generating capacity of wind farms has risen rapidly to 82 GW in the United States. With 487 GW, China is the country with the greatest wind power; many other nations, especially in Europe, are also actively pursuing wind power generation technology. The Europeans currently lead in offshore wind turbines, while the United States has favored onshore turbines. The problems and possibilities of the two types are different. Wind speeds are greater at sea, because the friction between moving air and the surface is less than on land (the power of a turbine increases rapidly with wind speed) and the winds tend to be more regular. On the other hand, onshore wind farms can be strategically placed on a ridge or in a valley to capture winds, and onshore installation and maintenance is much less expensive. Then again, offshore wind farms are less intrusive.

The US Department of Energy has big plans for wind power, and expects that some 20% of electricity generated in the United States in 2030 will come from the wind—this is an enormous generating capacity. It will be almost entirely concentrated in large wind farms and not in small rooftop turbines—the wind equivalent of solar panels attached to roofs—because of the scale effect for wind turbines. The only cloud on the horizon is the already-mentioned reduction in subsidies that is occurring at the present time. These reductions amount to an across-the-board decrease in the investment capital pumped into renewable technologies in many countries, and they are a product of the 2008 financial crisis and economic downturn. They are also a result of the successes achieved by renewable energy generators—particularly solar and wind—which are perceived by many policymakers as having reached their own tipping points, and so can now take off without further incentives.

Wind energy is clean—it generates no carbon dioxide—though it is considered by some to do environmental damage by killing birds, generating noise, and aesthetically degrading the landscape. Also, as we will see in chapter **26**, wind farms cover a lot of area. The technology is proven, though it is currently expensive; I can see wind power being a significant contributor to electricity generation in developed nations, perhaps amounting to 20% of world production by 2050.*

• • •

There are other renewable energy sources, though they are not so obviously good and will not become major sources of power in the decades to come or deeper into the Anthropocene. Biofuel and the three Ws—waste, waves, and wood—constitute some 11% of the world's energy production today, but this will likely fall, and for very good reasons. In a future when we will need to expand food production (see chapter **12**) it makes no sense to surrender valuable agricultural land to grow biofuel—corn for ethanol or soybeans for biodiesel oil. Biofuels are part of the renewable energy mix in some large countries (the United States, Brazil, and China, mainly) but the density of energy generated per unit area of land is low; we will see the importance of this notion in chapter **26**. The same applies to wood burning: the energy it gives us does not justify the space it takes up. Burning waste can play a small role, and a useful one if emissions are controlled, but it will not be a major contender in the energy stakes.

Wave power has been around since the 1980s, but there appear to be serious technical issues with implementing it commercially. The

---

*For an overview of wind power generation history and prospects, see Denny (2013) chapter 8. For current issues see Cardwell (2012), Gosden (2015), von Kaenel (2015), Magill (2015), and Ross (2015).

idea is sound—extracting the considerable energy of water waves in the littoral regions off coastlines and converting it to electricity, but the best design is not yet settled, and the wave power industry has long ago been overtaken by the sunshine industry and "windustry." It seems to me unlikely that wave power will perform more than a marginal role (pun).*

23 The Bad

Coal, oil, and gas are collectively known as "fossil fuels" because they are formed from decayed plant and animal matter that has been exposed to geological processes and turned into combustible material. Fossil fuels are bad—indeed, fossil fuel power plants are our second technological monster—because, of course, they accelerate (indeed, precipitate) global warming by generating vast amounts of atmospheric carbon dioxide. Organic matter is carbon based, by definition ("organic chemistry" is defined as the chemistry of carbon compounds) and so fossil fuels are mostly carbon. Coal is mainly elemental carbon; oil and natural gas are hydrocarbons—compounds made from carbon and hydrogen. Burning fossil fuels releases vast amounts of carbon dioxide; from elementary chemistry we can see, for example, that burning a pound of coal generates over three pounds of CO_2. Noting that  for each person on earth, 6.1 pounds of coal are burned each day, you see that coal contributes massively to global warming (a quarter of the world's energy comes from coal). For each American 18 pounds of coal are burned each day; thus you can see that the United States contributes disproportionately (half of America's electricity supply is generated by burning coal).

So burning coal warms the atmosphere as well as (historically) warming firesides. Coal is our dirtiest and cheapest fuel. Coal burn-

---

*See Ansolabehere et al. (2007) for the global energy contribution of biomass and waste. For more on biofuels, see Bourne (2007), the *Economist* (2013b), the *New York Times* website *Biofuels,* which contains links to many articles on different aspects of this subject, and the European Biofuels Technology Platform webpage "Global biofuels—an overview."

ing contributes more than any other fuel to the outdoor pollution that leads to more than three million premature deaths each year (Le-lieveld, 2015; see also chapter 8). It gets worse: when coal is burned, trapped radioactive particles are released into the atmosphere; in the United States, people who live next to a coal-fired power station re-ceive a greater dose of radiation than those living next to a nuclear power plant, if both are operating normally.*

Why use coal, given its dirtiness and huge contribution to global warming? The answer is that it is cheap and plentiful. Coal deposits are everywhere—three of the world's greatest coal users (China, India, and the United States) all have immense coal reserves. One day soon, when effective worldwide environmental regulations restrict the amount of CO_2 and pollution that utilities release to generate power (I am, just for this sentence, being an optimist) coal burning will die out, right? Nope—coal production in the United States is expected to stay fairly constant for the next three decades. Domestic use will decline due to cheap natural gas, but the slack will be taken up by exports, especially to China. The coal industry is in a state of flux today, after two centuries of dominating the growing energy market. Demand for dirty old coal is being stoked by the emerging economies (see chapter 7) and in the developed nations *clean coal* is being touted by the industry. Clean coal is the same product, but burning it results in much less carbon dioxide and pollutants such as sulfur dioxide and heavy metals released into the atmosphere, due to hi-tech combustion techniques and costly remedial and "premedial" measures taken at the power plant. These carbon capture and sequestration measures, if fully enacted, will remove maybe four-fifths of the emissions.†

---

*Even when nuclear power plants are not operating normally, coal can contribute more to radioactive contamination. The coal that was burned in the United States during 1982 released 155 times as much radioactivity into the atmosphere as did the accident at Three Mile Island 3 years previously, according to an Oak Ridge Na-tional Laboratory paper (Gabbard, 1993). The nuclear industry relishes the publi-cation of these kinds of data.

†Capturing the carbon dioxide before it is released, and burying it underground, is known as CCS—carbon capture and sequestration (or storage). There is a boatload of books and a raft of research dedicated to CCS alone, but here it generates just this single footnote, such is the breadth of our interests. CCS is being researched

The US coal industry is in a state of flux, despite the prediction of level output for the next 30 years, because of changing times and rising awareness of environmental concerns. The push toward clean coal and increasing demand from abroad means that by the end of the decade, coal will surpass oil as the dominant fossil fuel. The number of coal miners has fallen by 15% over the last 20 years, but at 120,000, it still constitutes a large body of registered voters. The center of US coal mining has shifted from the east to the Midwest, as old seams are worked out and new ones, once considered unprofitable to mine, are opened up. Given its long history, effective lobbying, cheapness (despite the cost of the remedial and premedial carbon capture and sequestration measures mentioned above), and abundance, dirty old coal is going to be with us for the foreseeable future.*

●　●　●

Oil tends to get itself into the news headlines more than coal, perhaps because it is associated with volatile (again, a pun) parts of the world, so its price is unstable. At the time of writing, oil is priced very low— this morning the Brent Crude spot price is US\$47.64 a barrel, down some 65% in a little over a year. This sharp fall has been due to Saudi Arabia maintaining high output, probably in a bid to crush the nascent US shale oil industry.

Whoa, let me back up the horses a little and provide some explanation. Oil currently accounts for a third of the world's energy use and about the same fraction of manmade atmospheric carbon dioxide. It drives industry and is a major geopolitical factor. Conventional crude oil—the stuff that is pumped out from under the Middle East or the North Sea, for example—is located in large reservoirs in only a few parts of the world. Extracting it economically and refining it to drive western automobiles and industry has been the achievement of the Seven Sisters—seven large private western oil companies that domi-

---

heavily in developed countries and also in China, a developing economy with infamous pollution problems.

*See Ansolabehere et al. (2007), Brink (2015), Denny (2013) chapter 4, Gabbard (1993), McBride et al. (1978), Miller and Smith (2014), and Nijhuis (2014a).

nated the industry for most of the twentieth century. In the last 30 years, these 7 companies have played a lesser role as new, state-run oil companies working for developing countries have taken over. Geopolitics and dwindling conventional reserves* have driven developed nations to find ways to extract oil that lies beneath their own soil. Technological advances and the high oil prices of recent years have, until recently, made it economically feasible to extract *unconventional oil*— the shale oil and tar sands oil that have become rapidly increasing components of our supply during the last decade.

The good news about unconventional oil is that there is a lot of it in, or rather under, our own backyards. The bad news is that it is harder to extract and process than conventional crude. *Kerogen* is a complex mixture of waxy organic compounds attached to certain shale formations; when extracted and heated, it turns into crude oil. Extracting the kerogen involves more than simply pumping: the shale is first hydraulically fractured, or *fracked,* to break it up. Some 60% of new oil and gas wells developed in recent years have required fracking. Fracking is controversial because it is feared by many to cause environmental damage, for example, by contaminating groundwater and causing mini-earthquakes; see chapter **24**. Shale oil is expensive to make. Plus, fracking and then heating the extracted kerogen takes energy, so the resulting oil contributes more per gallon to global warming than does a gallon of conventional oil.

Tar sands—such as the massive Athabasca oil sands in Alberta, Canada—contain *bitumen*, a mixture of heavy hydrocarbons that resembles cold molasses, but is not nearly as sweet (it is "sour," in oil industry slang, meaning it has a high sulfur content). In addition to heavy hydrocarbons, which can be "cracked" to form lighter, more volatile, and useful compounds such as octane and propane, bitumen contains a lot of impurities such as sulfur, nitrogen, and heavy metals. The bitumen is expensive to extract—expensive in energy as well

---

Peak oil is the year when oil extraction reaches its greatest extent—by definition, it dwindles thereafter. The year of peak oil is often taken to be the beginning of the end for the industry as a whole. Obviously there is much speculation about when this will be, or when it was. For a recent article suggesting that it might be anytime now, see Patterson (2015).

as money—and so, like shale oil, it is worse for the environment and climate than conventional oil. Such is the size of Canadian tar sands deposits, however, that Canada is now the source of more oil imported to the United States than any other country—contrast this fact with the dependence of US oil supplies upon Middle East sources in the 1970s, and you can see how the industry has changed. Obviously, transporting oil from Canada is easier and cheaper (though the proposed Keystone XL and Northern Gateway pipelines are drawing opposition from environmentalists), plus Canada is more politically stable and friendly, so supplies are more reliable.*

● ● ●

If developing unconventional oil seems like scraping the barrel, then how about the least offensive of the fossil fuels—natural gas? Gas accounts for about a fifth of the world's energy usage and a fifth of atmospheric CO_2 generated by humans. In the early days of oil production, associated natural gas was burned off at the wellhead as an unwanted byproduct. (There is also "unassociated" natural gas, found in fields where there is no oil.) Now, however, we recognize its value. The stuff that comes out of the ground is a mixture of light hydrocarbons, which are purified to create methane before being pumped to customers. Natural gas is easier to transport than oil, and it burns more cleanly. Worldwide, there are greater reserves of natural gas than of oil.

Conventional natural gas sourcing is as problematic for western countries as is oil sourcing: the two countries with the greatest proven reserves are Russia and Iran. However, shale gas is plentiful in the United States, Canada, and Europe. Shale gas is unconventional natural gas that is obtained from oil shale deposits; it also requires fracking the shale but thereafter is easier to extract and process. Shale gas production has risen rapidly over the last 12 years and now accounts for half of US natural gas. If humans have to burn fossil fuels—and currently we must, and will for decades to come (see chapter 22)—then

*For more on conventional and unconventional oil history and development, see, e.g., Biello (2013), Chase and Jones (2015), Denny (2013) chapter 5, Goldenberg (2014), McCarthy, Chase, and Jang (2014), Patterson (2015), and Patton (2015).

gas is the least bad form to utilize. Even so, it has a limited supply—a few decades worth, depending on world population and economic growth (the first of which we can predict sensibly, the second we cannot)—and it is not renewable.*

24 Frack Attack

Hydraulic fracturing is the key to extracting unconventional oil and gas from shale deposits. Shale oil and gas could not be extracted economically without fracking; fracking is technically difficult and expensive. Put these facts together, along with the explosion (an unfortunate word choice—let me change it to "rapid rise in") shale gas output in the United States, and you immediately deduce two things about the present-day energy business: (1) the technology of fracking has now been solved, and (2) oil prices have been high enough to justify the pursuit of this technology.

One of the positive features of oil shale rock formations is that they are all over the place—not just underneath land that is part of hostile nations in volatile parts of the world—so unconventional energy sourcing is secure. Fracking is performed in the United Kingdom, in the United States, and in most of the developed nations that need the oil and gas fracking produces. The technical difficulty is in part due to the need to drill vertically down to a shale deposit and then twist the drill bit so it drills horizontally within the shale—a neat trick. The fracking part consists of pumping water (with some sand and chemicals mixed in) into the bore hole under high pressure. This action fractures the shale and releases the oil or gas, which is forced up through the well head. The success of the new technology can be seen from the effects on US domestic gas and oil prices (down) and production (up). A spokesman for Trafegura, a Dutch commodity trading company, said "growth in U.S. shale production has turned the distillates

*For an upbeat assessment of recoverable shale gas reserves, see the US Energy Information Administration website at www.eia.gov/analysis/studies/worldshalegas. For a contrary view see Orcutt (2014) and Powers (2014).

market on its head" (Yep, 2014). No wonder the price of oil and gas has dropped: not only are the Saudis trying to flood the market with cheap conventional oil and so kill the more expensive unconventional fuel industry, but the very success of that industry is itself driving prices down. Of course, this situation is unstable and will soon change, but how—in which direction?

Fracking has come under attack by environmentalists because it causes damage: it may contaminate local groundwater, and it causes minor earth tremors (thus, magnitude 1.5 and 2.2 tremors due to fracking were recorded near the city of Blackpool, England, in 2011). Environmental concerns have led to scrutiny of fracking practices, and to various degrees of regulation and restrictions in different countries. Thus European Union countries are in the process of setting up regulations but, characteristically, the United States is not. Lax US regulations such as the "Halliburton loophole," which exempts the oil and gas industry from the requirements of the Safe Drinking Water Act, has created local environmental degradation and human health issues wherever large oil shale deposits are being fracked, such as North Dakota and Texas. Without federal restrictions, states (and communities within states) are coming up with their own rules governing fracking; Vermont and New York have banned the practice.*

㉕ The Ugly

Physicists tell us that there are four fundamental forces in the universe. In order of increasing strength these are: gravitational, weak, electromagnetic, and strong. We are familiar with two of these forces because they extend over long ranges—long enough to encompass the human scale. The force of gravity is so weak that it is felt only on very large scales between large objects—planets, or galaxies, or between a person falling out of a window and a planet. The electromagnetic (EM) force is much, much stronger, as you can see when you hold a small magnet and use it to pick up a paperclip; the EM force of the

---

*For informative popular articles on fracking, see BBC News (2013), Dobb (2013), Goldenberg (2014), Tran (2014) and Yep (2014).

magnet acting on the paperclip overcomes the gravitational force of the entire Earth. The other two forces are not familiar because they reside inside the nuclei of atoms, and are so short range that their effects are not felt outside these nuclei. The weak force need not trouble us further here, because it does not contribute directly to the power needs of humanity; the strong force is responsible for nuclear power.

Hydropower is underwritten by the force of gravity; hydro projects are very large because gravity is so weak. Almost all the other sources of power are underwritten by the EM force: induction motors, electric batteries, oil, coal, gas, biofuel, muscle power. Anything that is directly powered by electricity, magnetism, or chemical reactions relies upon the electromagnetic force.

The second law of thermodynamics (that the total entropy of an isolated system always increases over time) is profound, and it applies everywhere. One of its consequences, of relevance to us here in this account of power generation, is that there is no such thing as perfect efficiency. Thus, a perfect refrigerator would remove heat from the inside and dump exactly the same amount of heat on the outside, but in practice there is always more heat on the outside. A perfect gas-fired power plant would deliver all the chemical energy stored in the gas molecules—but in the real world there is always waste heat, as well as byproducts such as carbon dioxide. A perfect horse would exert all the energy it took in as food to pull wagons or transport humans, but real horses produce a lot of horseshit—a superfluity of equine ordure. The disposal of this waste product in large cities such as nineteenth-century London was a major problem; waste disposal issues are not new. For nuclear power plants, the obnoxious waste product is radioactivity.

● ● ●

Nuclear power emerged in the 1950s as one of the bright hopes for the future of humanity, following the global disaster that was World War II. The fuel for nuclear power plants is a natural substance—uranium ore, which is processed ("enriched") to increase the concentration of uranium-235, abbreviated ^{235}U. When this uranium isotope is bombarded with slow neutrons (low-energy neutron particles) it splits or *fissions* to produce radioactive byproducts and a lot of energy. The energy content of ^{235}U is so great (because the strong force is 137 times more powerful than the EM force) that it almost doesn't matter how much of the ore there is in the world—we can, for all practical purposes, say there is enough uranium ore in the surface and oceans of the world to last forever, so that nuclear power is effectively a renewable energy source. Also, nuclear power has a nonexistent carbon footprint—it generates no greenhouse gases and contributes nothing to global warming, though it is a potential threat to the environment. In the optimistic early days of nuclear power Lewis Strauss, head of the US Atomic Energy Commission, implied that electricity supplied by nuclear power would be "too cheap to meter" (as reported in the *New York Times*, September 17, 1954).

Strauss and the nuclear industry he represented were wrong about the cost of nuclear power, because nuclear power plants are phenomenally expensive to build, and the safety procedures they follow (including waste disposal) are also very costly. Opponents of nuclear power are wrong about the risks it poses (see chapter **28**). There is much disagreement—even the "too cheap to meter" quote is contested. Thus, opponents dispute the endless fuel supply. Proponents point to the low body count resulting from nuclear power compared to any other fuel source, even hydro, but opponents dispute this claim. Opponents point to the risks of radioactive contamination, while proponents provide statistics to dispute this. There are few neutrals in the nuclear debate. As a result of this debate and of power plant accidents, the status of nuclear power plummeted over 30 years from The Goose That Laid the Golden Egg to The Ugly Duckling of power generation. Nuclear power is our third toolmythological monster, because for the last 50 years, it has been perceived as a sword of Damocles, ready to smite us for our hubris—we are playing with forces we don't understand.

There are different types of nuclear reactors supplying the power in nuclear power plants. Some use plutonium or thorium isotopes for fuel instead of uranium. Different kinds of coolant (light water, heavy water, molten salt, molten metals) and different reactor designs exist. The most common design type is the PWR (pressurized water reactor) favored by western nations; the most infamous is the Soviet RBMK (which translates as "high power channel-type") reactor. Most experts consider the design of modern nuclear reactors to be inherently safe, and the design of the RBMKs to be inherently unsafe. Most reactors today are of *once-through* design, in which the fuel is used once and then discarded (how to discard it is a problem). *Fast-breeder reactors* process this spent fuel into forms that can be reused as fuel. The nuclear fuel recycling idea is, like everything else in this troubled industry, fraught with controversy. Nuclear power contributes about 20% of current US electrical power needs, and about 11% of world electricity (i.e., 3% of total power consumption).*

●　●　●

The 1979 accident at the Three Mile Island nuclear reactor in Pennsylvania resulted in radioactive coolant leaking out. It was caused by mechanical failure, compounded by human error; its main effect was to put the US nuclear power program on hold for 30 years. Most US reactors are now over 30 years old, and today more are being dismantled than commissioned. The much more serious accident at Chernobyl in Ukraine in 1986 released radioactive material into the atmosphere, promptly killing 31 people and ultimately leading to several thousand fatalities due to diseases caused by radiation. This accident was due to reactor design flaws and operator error. As a result of these accidents, the nuclear power industry worldwide has been languishing for the last three decades.

The huge initial price tag of a nuclear power plant—plus the penchant of the industry to invariably underestimate construction costs and timescales—leads to doubts about nuclear power's cost effectiveness—it is today still more expensive than coal or gas, though much less expensive than other renewables. But it is safety, not cost, that first comes to mind when people express concerns about the industry. There are 46,000 tons of low-level nuclear waste sitting in temporary sites around the 99 US nuclear power plants because the government dithers over building a permanent waste storage facility. There is no permanent facility anywhere in the world for storing high-level waste. (High-level waste consists of 3% of the total waste from nuclear reactors and contains 95% of the radioactivity.)

New nuclear power plants are being built in the Middle East, in Korea, in India, and, especially, in China. Thirty-two new reactors are

*For a survey of nuclear power and a description of the most common nuclear reactor types, see Denny (2013) chapter 7. For a fuller description of modern (Generation III) reactors, see the World Nuclear Association webpage "Advanced Nuclear Power Reactors."

being built there as I write, and plans for 200 more are in the works. New, safer, and more efficient designs led to a reevaluation of nuclear power in western countries a decade ago, and many new plants were proposed. Then came the world financial crisis of 2008 and the Japanese Fukushima-Daiichi accident of 2011. The accident was caused by a tsunami and a design fault. The design fault was not in the reactors, but in the decision to place them next to the coast (in a part of the world that is hit by a lot of tsunamis). A 15-meter wave generated by an offshore earthquake took out reactor cooling pumps, which led to the meltdown of 3 reactor cores and the release of a large amount of radiation. The aftermath worldwide included several countries deciding to phase out nuclear power—thus Germany has said it will decommission all its nuclear reactors by 2022 and will replace them with other renewable sources of power. At the time of writing, only 1 of Japan's 44 reactors is back online (all were shut down after Fukushima for safety inspections and upgrades). The nuclear share of world electrical power generation has dropped from 17% to 11% over the past 20 years, and the Fukushima disaster will only promote this downward shift.

The trend in developed countries and the contrary trend in emerging Asian nations mean that, according to the International Energy Agency, nuclear power will play a significant but not dominant role by the year 2040. It will provide less power than coal, less than gas, less than other renewables such as wind, solar, and hydro. Only oil, of the current major players, will provide a lower percentage of world power (because it will have run out or be too costly to extract).*

•   •   •

I will be arguing for a greater role to be played by the nuclear sector in generating electricity for the world by mid-century and certainly by the end of the century. Doubtless many readers are not quite ready yet for such an argument, so I will delay it or, rather, spread it out over several chapters—chip away at your nuclear prejudices. Undoubtedly the nuclear industry needs to get its act together to demonstrate its

---

*For a snapshot of the future of nuclear power, see Anderson (2015), Babbin (2015), Lecompte (2014), the *Economist* (2013c), and Prince and Shoulak (2015). These newspaper and magazine articles are ostensibly unbiased—not written by environmentalist or nuclear organizations.

comparative safety, develop permanent storage facilities (and perhaps develop fast-breeder technology and a secure method of transporting nuclear waste to and from reprocessing plants), and drive down the cost of nuclear-generated electricity. There are advantages of nuclear energy that currently keep it in the loop and make things difficult for those contemplating removing it entirely as a primary energy provider. It provides a regular source of power—not intermittent, like solar or wind power. Utilities like this characteristic; it makes it easier for them to provide a steady and reliable supply. Nuclear power plants can be placed on any geologically stable ground, unlike hydro or other renewables, and fuel sourcing and extraction is less problematic than for fossil fuels.

There is a lingering sense of disappointment with the nuclear industry because of the way it has quite spectacularly blown its brownie points over the decades. In the 1950s, nuclear technology was seen in the United States as having contributed to ending the worst war in human history—though perhaps in the mind of Joe and Jane Average, the association of nuclear bombs with nuclear power was an unfortunate if unjustified consequence. Science fiction movies from that time often portrayed the hero as a nuclear physicist—with tweed jacket, pipe, and other academic accoutrements—who saved the world and got the girl. Today, a nuclear physicist is portrayed as an otherworldly nerd at best or a sinister megalomaniacal mad scientist at worst. I suspect that to reverse this trend—to rebrand nuclear power as an acceptable member of the brood of power generators and not the ugly duckling at the end of the line—the nuclear industry will need to do more than simply educate the Averages, Otto Normalverbraucher, and others. It would be uplifting to think that education on its own might be sufficient, but I doubt it would be. There are plenty of educational websites out there, provided by the nuclear industry as well as by those who oppose it, but it appears to me that only converts read them. Sadly, it seems that education will need to be fused with spin to get the attention of a public jaded by the energy debate and inundated by online information and misinformation. Perhaps an attractive woman in a short skirt, or a muscular dude with his baseball cap on backward, could represent the nuclear industry in a reality TV show. Am I being too cynical?

Talking of fusion—and this is the worst segue in the book, I promise—what about the future of *fusion* reactors? All the nuclear reactors in the world today are *fission* reactors, in which nuclei of uranium or some other heavy element are bombarded with neutrons and split, releasing energy. In nuclear fusion reactions, the nuclei of light elements such as isotopes of hydrogen are slammed together and stick (hence "fusion"), which also releases energy. The energy release is even greater than that of fission reactions, and there are no radioactive waste products. The fuel is heavy hydrogen, which is not expensive to produce. Fusion reactors are the ideal way to generate almost limitless clean power—in an ideal world. In the real world, it has proved all but impossible to contain the fusion reaction, and so commercial fusion reactors are a distant dream. They exist in laboratories, however, though the lab reactors consume far more energy than they produce. I remember talking to a young nuclear physicist 30 years ago about the prospects for commercial fusion reactors (he was based at the Culham Laboratory in Oxfordshire, England, which boasted a large—for the day—*tokamak* fusion reactor); he said, "not in my lifetime."

Perhaps in the year 2100 humans will have developed commercial nuclear reactors and clean nuclear power that is too cheap to meter—if so, they will be fusion reactors, and they won't be here much before then.*

26 Gridlock

In this chapter I look into present and future electricity grids—the network of power lines that snakes across our nation and every other to a greater or lesser degree, depending on level of development and infrastructure spending. These power grids connect generators with domestic and commercial consumers. Let me begin by getting all the dreadful electrical puns out of the way, in one brief summary paragraph.

The current electricity grid system has the potential to transform the way in which power is moved around a network and around the world, if the induction to change is not matched by too much resistance. In the developed world, many national grids are aging, which impedes the changes needed to meet the increased capacitance that is

*For the prospects of fusion, see Denny (2013) and Jha (2015).

anticipated down the line. The problem is cost; the gain is increased power supply stability, efficiency, and security.

• • •

There is a lot more to a grid than the power lines that connect supply with demand, however. Transformers enable cheap and efficient transmission over long distances, and distribution and control networks attempt—usually but not always successfully—to keep a lid on the inherent instabilities of such directed, complex networks. Let's begin with a basic description. The simplest grid structure would consist of power lines radiating out from a generating plant directly to its consumers, but this is not how things are usually done in practice. An important aspect of electricity supply is its steadiness, or quality (constant voltage and frequency), and supply quality is difficult if there is only one generator; if it goes down for whatever reason, then supply is completely lost. In major networks the generators and consumers are meshed together, interconnected so that if one generating plant is knocked out, or one power line goes down, supply is obtained from other plants and is transmitted via other routes. All this happens— most of the time—near the speed of light.

Physics dictates many of the transmission line characteristics. Resistance to flow is less, for a given power, if the voltage is high and the current low. So step-up transformers raise the voltage (a.k.a. potential, or tension) outside power plants so that the electrical power can be transmitted for long distances at high voltage. The power loss— dissipated as heat—increases with line length, so it is important to maintain high voltage over as long a distance as possible. (Put another way, power loss per mile of line is reduced if the voltage is high; but power loss also increases with line length.) Close to the consumer, step-down transformers reduce the voltage. The numbers vary: turbines in generating plants churn out electrical power at voltages of 10–20 kV, which is cranked up to either 230 kV or 400 kV in Europe, 500 kV in most of the Americas (and up to 765 kV in Canada), and as much as 1,100 kV in China. For the end user, potential is stepped down to 220– 240 V in Europe and 110 V in North America. Voltage varies in other parts of the world (Japan uses both 220–240 V and 110 V).

For various reasons that have to do with end use as well as generation, the AC frequency that is usually adopted by most countries is 60

Hz. Within each grid—no matter how widespread it is and how many different types of turbines pump power into it—the power must be synchronized, i.e., have the same phase. Phases vary from one grid to the next, but phase within a grid is carefully maintained.* Sometimes it is necessary to transmit power from one grid to another: this transfer cannot be done as AC power because the phases are not matched; the result would be a disastrous fluctuation of power and increase in power loss. So when power is transmitted across grids, it is done as HVDC (high voltage direct current). DC switching and transforming are much more expensive and difficult than the equivalent AC operations, and are currently an area of active research. Consequently, it is economical to introduce HVDC lines only over long distances, exceeding 500 km. In North America, there are three main power grids connected by seven HVDC transmission lines.

Long-distance transmission of electrical power is cheap (losses are typically 2%). Added to transmission loss is distribution loss, which varies from region to region because of differing hardware components, grid organization, and culture (corruption and "power pilfering" are rampant in some countries). Thus, in Japan the power that is lost between generator and consumer, due to transmission and distribution, is 5% of the total. In the United States it is 6.5%; in Great Britain it is 8.9%; in India it is 27%.

Maintaining the stability of electrical power is a dynamic exercise; it is not just a matter of keeping a steady supply and monitoring a few loads at a few points across the grid. There is a reasonable analogy here with the domestic household phenomenon of water hammer—the knocking and shuddering that occurs in some water pipes when you switch a faucet on or off. Water hammer is a pressure surge that charges around the network of pipes in your house, bouncing off valves and sharp bends. It can cause serious damage, bursting pipes and plumbing attachments, so special components are installed to minimize the chance of water hammer occurring and minimize the damage it does. Valves are designed so flow rate changes slowly, not suddenly; arrestors, ac-

*For readers who, like me, happen to be science nerds, I note that in fact there are three phases needed, differing by 120°, to guarantee a high-quality electricity supply. The three phases are synchronized across a grid, but differ by a random overall phase factor from one grid to the next.

cumulators, and perhaps expansion tanks are inserted in the pipeline network at strategic points, to discourage or dampen any pressure waves that develop. In an electrical grid, power surges replace
water pressure surges, lines replace pipes, etc.

Consider the situation from the point of view of a utility company.
Its job is not usually to generate electrical power (that is often done by
a different company) but to transmit the power smoothly to the end
users. A network of lines—the grid for which the utility company is
responsible—is being pumped full of electrical power by a number of
generating plants; this power must be expended by the end users at
the same rate. If power accumulates in the grid something will blow,
and if it is dissipated due to losses then the consumers experience a
brownout or blackout. Power darts around the network at lightning
speed, controlled—but barely, sometimes, like a caged beast—by the
utility. (Here's another analogy: imagine a juggler spinning plates on
poles, with a pack of dogs fighting each other between the poles.) Demand varies as consumers in one region switch on air conditioners during a sweltering day, or switch on kettles at halftime during a major
televised sporting event. Demand falls at night, but many grids cross
time zones, so lights-out time varies across the grid. Overhead power
lines sag due to heat expansion, or due to the weight of accumulated
ice in some regions, and droop into trees, causing short circuits that
break the line (for reasons of cost and weight, overhead lines are un-
insulated). Trees fall across lines, or trucks slam into poles. For any
number of reasons, the network can suffer breaks that have to be compensated for, with power rerouted along different lines to maintain a
steady supply to the end user. Overhead power lines are much cheaper
to install than underground lines but are more expensive to maintain
because they are more vulnerable, and so teams of linesmen are made
available 24/7 by utility companies to repair damage to overhead lines.

The grid must talk to the source—to the suppliers of power. When
end-user demand varies, the generating plants must respond by changing the power they pump into the grid. This can be done quickly for
some technologies but not others. Thus, for example, if a hydro plant
is asked to reduce power output, it can divert some of its power to
pumping water uphill into storage reservoirs, where it is kept until
needed. During times of increased demand, this water is then released
to power extra turbines. Other technologies, such as solar, store power

for short periods in batteries or capacitors; otherwise they must shut down when demand is low.

Many developed nations installed their electrical power transmission and distribution infrastructure decades ago—in the United States, much of the equipment dates from the '60s and '70s. Stress this aging network with increasing power demand, and you can expect—and get—increasing system failures. The big 2003 power outage that occurred in the Northeast, the Midwest, and regions of central Canada was due in part to sagging lines and in part to an inadequate response by a utility company. Fifty million people lost power, and 500 generators shut down. The damage spread in a domino effect across the grid, as the system failed catastrophically.

Grids are inherently unstable, and outages are prevented by good design and operation, by carefully monitoring the system and quickly responding to perturbations. In developed countries the result is a reasonably reliable electrical power supply—thus in Japan consumers lose an average of 4 minutes supply per year. In the United States, for a number of reasons, losses are greater: 92 minutes per year in the Midwest and as much as 214 minutes per year in the Pacific Northwest. These losses exclude those due to extreme weather events, and can be attributed mostly to aging infrastructure and lack of investment.

The number of power blackouts has increased in recent decades, partly for these reasons but partially because of an increase in extreme weather events. Federal data show that there are 285% more large-scale blackouts than there were in 1984. Each time such an outage occurs, the cost runs from tens of millions to thousands of millions. The trends for the Anthropocene are not encouraging. Over the 40 years from the '50s to the '80s, 2 to 5 major power outages occurred each year in the United States. In 2008 the number increased to 70, and in 2012 it was 130. Extreme weather rose from being a minor cause of such outages (20%) to being the major cause (70%).

I count the aging electrical power plants as our fourth technological monster. The power of the generating plants rattles the electrical grid cage, and threatens to break out sometimes, causing power blackouts. Meanwhile, the power plants belch forth much of the anthropogenic carbon dioxide that is leading to climate change. The monster is partly hidden under the bed, like all the best monsters: some of us drive electric cars so we can feel good about our carbon footprint, but we know in the back of our minds that most of the electrical power in our city comes from coal-fired power plants.

Everybody within the industry, and many consumers sitting in the dark at home, know that a major upgrade to the power grids is needed. Again focusing on the United States (though many other developed countries are in the same boat) I note that the electrical power supply system consists of some 7,000 generators connected to end users by 5 million miles of power lines. The whole shebang is valued at $876 billion. The country needs—some would say desperately—an upgrade to a modernized smart grid system, costed by the Electrical Power Research Institute at a staggering $338–$476 billion. To seed this process, in 2009 the federal government injected $4.5 billion. This smart grid, if developed and completed, would ensure a more stable and adaptable supply even as demand grows, more efficient use of electrical power, more security of power supply in the face of terrorist attacks, and better integration of power from renewable sources—of which more shortly. Before discussing the impact of renewable power sources, however, I need to bring you up to speed on the notions of *power area density* and *distributed power supply*; it sounds tedious but I will be brief—and it is important, so suck it up.*

• • •

Vaclav Smil,† a respected energy expert, came up with an interesting and perhaps neglected aspect of the energy/power generation industry a few years ago, which casts an illuminating sidelight on the different technologies. He asks for the geographical area taken up by a power plant and performs detailed calculations to obtain the answer. The result is a number—the number of watts generated per square kilometer of plant area—that can be compared across technologies, with results that can and should influence the development of some re-

---

*A more detailed summary of electrical power generation, transmission, and distribution is provided by Denny (2013) chapter 3. For more on the aging US grid infrastructure, see M. Clark (2014) and Tollefson (2013).
†Smil is a Czech-Canadian professor at the University of Manitoba; he has published many papers and popular articles on the energy industry and in particular on the power area density idea. See Smil (2015).

newable energy sources. These calculations are not necessarily straightforward. Thus, to estimate the power area density of the coal industry, Smil considers not just the energy contained in coal and the area of the mine it comes from, but the size of the tailings ponds, and the size of the power plants that burn coal, including storage yards, fly-ash disposal sites, etc. For nuclear power, the size of uranium mines, reprocessing or heavy water plants (if used), waste storage sites, etc., must be considered, as well as the reactor site area. Let's begin with a simple example. Wood grown as biofuel generates about 10 MWh (megawatt-hours) of energy per hectare of land per year. This energy density corresponds to a power generation capability (assuming no wastage or inefficiency) of about 114 kW/km^2 so, if wood biofuel were to be adopted as a major source of energy for the world—say enough to generate a terawatt of power (about 5% of current global needs)—we would need to grow it on 8.8 million square kilometers of land—more if the process of converting the wood energy into electricity is less than 100% efficient. This amounts to a lot of land (it is equivalent to a square that is 1,840 miles on each side), which would be better used for growing crops to feed an increasing population—see chapter 12. Biodiesel energy density is four times greater than that of wood; even so, it is still much too low (and so requires too much agricultural land area) to be a major source of power.

Smil estimates that the power area densities for coal and gas are higher than for any other source; this is one reason why these fossil fuels have dominated large scale power generation. Nuclear power also has a high energy density, but the other renewables are much lower. Thus for hydro, the area taken by the upstream reservoir of a dam must be factored into the calculation. Proponents of hydropower might argue that, for example, the area taken up by Lake Mead (formed by the Hoover Dam) does not impact society negatively, because the desert valley that used to be there was unpopulated and of no economic value. The same cannot be said of the Three Gorges Dam in China— over a million people were relocated (they had no choice in the matter) because they lived on land that would be covered by the dam headwaters.

Using Smil's numbers for photovoltaic (PV) solar power (he estimates a power area density of between 4 and 9 W/m^2), we see that if it needs to generate 20% of global power by 2050, sunshine farms will occupy about 540,000 square kilometers of land by mid-century. This figure is much less onerous than the biofuel area, and much of the land needed can be dry desert, which is inhospitable for humans and agriculture alike. Also, an important advantage of PV solar technology is that it can be small scale: panels can be put on urban rooftops, which does not take away land area (though firemen don't like the idea). On the other hand, generating such a high fraction of global power in this manner will require some 100 billion solar panels (assuming 20% panel efficiency), which will constitute a significant environmental challenge in terms of production of new panels and disposal of old ones.

Wind farm power area density is tiny, according to Smil (between 0.5 and 1.5 W/m^2); from his figure we see that, if the United States is indeed going to generate 20% of its power via wind technology by 2030, it will have to give up a quarter million square kilometers (60 million acres—a little under a hundred thousand square miles) to this end. Much of this acreage may be used simultaneously for crops or grazing. Even so, it is clear that wind turbines are going to be a common sight for many Americans in 15 years' time—unless, perhaps, the United States develops an interest in offshore wind generators, which it has recently shown signs of doing.

So Smil's power area density concept tells us much about the room to be taken up by renewable energy technologies over the coming decades, though the impacts on humans of this extra space requirement (extra compared with the area required by our current fossil fuel sources) varies with the technology. Renewables require much more space, so it will be necessary to develop new and extended infrastructure for electricity distribution, just because of the extra land area involved. I will now show you why this and other powerful reasons suggest a new grid infrastructure.

• • •

The greater area of land that will be taken up by renewable sources of energy, compared with current sources, has wide implications for power grids. Renewable power technologies are sometimes described as "distributed," because of their spatial spread. In particular, they may spread into existing space in small units (as with rooftop solar panels) as well as expand to large-scale power plants. The point is, the power feeding into the grid by mid-century will be from tens of thousands of locations, not just a few dozen zip codes that include large power plants. Rooftop power will supply the house beneath it but, when the house is drawing little power, the supply can be input to the grid (given smart meter and smart grid technology) so that many small-scale and intermittent suppliers will be mixed in with more traditional large-scale power plants in the future. Couple this characteristic of distributed power with the fact that most forms of renewable energy are inherently intermittent (solar power and wind power, for sure) and you see that the management and control of power grids in the future is going to be an order of magnitude more complicated than it is at present.

So: managers of power grids in the decades to come will have to contend with variable supply as well as variable demand. Thus, if solar power is part of the grid, output will fall when a cloud obscures the sun; output will decrease across the Texas Interconnection (one of the US grids) when the wind dies down in West Texas—Texas is big on wind farms. A future grid will extend into deserts and onto city rooftops, to offshore wind turbines and suburban carports. Supply will vary when Mr. Smith unhooks his electric car from the grid in the morning and goes to work. He will have a contract with his local utility to supply them with power from his car battery overnight, when he doesn't need it. They will pay him at a certain rate, and draw electric power from his car battery while ensuring that it is fully charged by 6:30 a.m.* Multiply by all the Mr. and Ms. Smiths in a utility's region of operation, and you can see how two aspects of electrical power supply are going to change in the decades ahead: supply will be much more variable in both type and size and, to harness it, grids will be much more extensive and flexible.

Flexible on the large scale, as well as within a suburb. In 2050, power from offshore and onshore wind farms, from hydro plants up in the mountains, from desert solar power plants, from nuclear power plants built in the middle of nowhere to allay safety fears, from some remnant gas-fired power stations that are about to be decommissioned because of their large carbon footprints, from household carports and

*The notion of utility power drawn from electric car batteries is called vehicle-to-grid or V2G; the negotiations are sometimes called "carbitrage"—see Denny (2013) chapter 3.

rooftop solar panels—all will have to be integrated to provide synchronized AC electric power to a nation. A smart grid will control the flow of both electricity and money. Your washing machine may be set to come on at night, when the price of electricity is lowest, as decided by the utility's smart meter in your house. Rapid switching will maintain a stable power supply, adding or subtracting power from steady sources such as nuclear or hydropower plants to counter the natural fluctuations from unsteady solar or wind farms. A smart grid will cost billions, but the result will be more secure and more reliable electrical power, because the grid is more organized and more intelligent and because it is larger, with a more varied mix of technologies that generate the power. Outages will be less common and less costly (currently $800 billion per year in the United States) and, because renewables will figure more prominently in the mix, carbon emissions will be significantly reduced.

Size matters, for electrical power grids—bigger is better. A sign of things to come: there are negotiations between nations about supplying power across borders, for what may turn into a World Wide Web of power—a super-grid. The idea of energy crossing borders is not new: after all, for years now Russia has been selling gas to western Europe, Canada has been exporting hydropower and fossil fuels to the United States, and so on. The changes in future will be a matter of degree. Europe will import electricity that is made from North African sunshine. The ever-present winds of the Gobi Desert will be harvested for power that is exported long distances (perhaps via superconducting HVDC connectors) to China, Japan, Korea, and Russia. Wind power from the North Sea and the Baltic Sea will supply the North Sea Offshore Grid. There are plans for transferring electrical power across borders via an Asian super-grid and a Southeast Asia super-grid (the latter connecting Australia, Indochina, Indonesia, and the Philippines). Some of the advantages of scaling up are obvious: unused capacity in one country (say, after dark) may be consumed in another country, in a different time zone.

An obstacle for the smart grids in our future—one that argues for large-scale super-grids—is the increasing effects of extreme weather. Weather is the greatest single cause of power outages, as we have seen, and global warming is leading to an increase in the number of extreme weather events. Distributed smart grids will be better able to deal with this problem, because the grids will be both more robust and more spatially extended.

The greatest hurdle to the implementation of super-grids will be cost—no surprise. But there will also be significant issues related to the negotiations between nations. It will be necessary to extend rules for trading electricity. Nations will have to agree on how cross-border hardware construction, installation, and maintenance is to be paid for, and it will be necessary to establish international codes of practice and standards.*

• • •

Bad old fossil fuels will dwindle, but slowly—we will still be using gas if not oil mid-century, and so will still be heating up the atmosphere well beyond 2°C, because fossil fuels are cheap and don't take up much space. Good renewables will increase, technology and common sense willing, but will not be able to fully replace fossil fuels—that is, to take up all of the slack.† Perhaps half of our energy needs must come from the ugly monster, nuclear fission, at least until the end of the century when (again, if we can develop both the technology and the common sense) nuclear fusion will see us into a safer and cleaner energy future.

I need you to confront our four monsters. Then I will emphasize one of them—nuclear power—because it needs to be buttered up to be made palatable. Then we're done with energy and it will be time to move on to the other side of the coin, the other half of the double act that can secure our future: common sense. We will need both technology and common sense in the Anthropocene.

---

*Interesting articles and books that discuss in more detail current electrical power grids, upgrading to smart grids and super-grids, include Achenbach (2010), Borbely and Kreider (2001), M. Clark (2014), Gellings (2015), Martin et al. (2013), and McElfresh (2015). See also the useful Energy.gov web article *Top 9 Things You Didn't Know about America's Power Grid.*

†There is a technical pun here—if you see it already, please award yourself a brownie point and move to the top of the class. Hint: in the coal industry, the fragments and dust that remain after screening.

27 Not *Monsters, Inc.,* nor the Four Horsemen

There are, in popular perception and reality, technological monsters out there. I have mentioned four of them as they rose their ugly heads in earlier chapters. They are (1) the agricultural industry, which these days includes GM (genetically modified) crops, (2) fossil fuels, (3) nuclear power, and (4) electrical power plants. Why these four, and why such an emotive label?

In terms of human creations—products of science and technology—these four are crucial for both understanding the Anthropocene and for helping us (I will argue) get through it with a reasonable quality of life for humankind. The agricultural industry can make a very good case for having saved billions of people from starvation (see chapter 12), but it has led to significant pollution, and has left us overpopulated and dependent on further progress in agricultural technology to avoid a bigger global famine. One promising advance and a significant aspect of the solution to this looming human disaster—GM crops, which will have increased yields and will be more resistant to disease—is itself a problem, in the eyes of many people. By messing with genetics we are uncorking a genie we won't be able to put back in the bottle; we are playing with forces we don't yet sufficiently understand. This is the kind of Frankenstein monster problem that has been discussed by Bruno Latour in a thoughtful article that appeared in a recent ecological compilation of pin-pricking essays aimed at shaking up some established environmentalist views. But I am getting ahead of myself—Latour's Frankenstein will walk the Earth, in the section after the ornament.

What of the other three technologies—how do they merit monster status? About fossil fuels little more need be said: they have powered our own industrial revolutions and are doing the same in many other countries right now, but they are responsible for the growing levels of atmospheric carbon dioxide that have led to global warming in our recent past, a warming that will continue for decades at least, if we do nothing to stop it (see chapter 16). Again, we do not sufficiently understand climate processes to take this action responsibly, and we do not know all of the consequences of what we are doing. Nuclear power: a similar story of hubris. We arrogantly unleash forces of nature that we may not be able to control. What about power plants? This technology—I should make it plural—these technologies are in some ways a combination of the others (some electrical power plants are nuclear; others are powered by fossil fuels to the extent that electricity generation is the biggest single source of greenhouse gases) but also add a new dimension to the human folly of runaway technology. We play the fiddle while Rome burns, consuming electrical power like there was no tomorrow (which, some doomsayers tell us, there won't be because of our overconsumption). We enjoy the benefits of abundant electrical power while we distance ourselves from the unpleasant aspects of power generation technology, in the same way as we stuff ourselves with steaks while avoiding slaughterhouses. The consequences of electrical power generation—producing air pollution that silently kills millions, while accelerating climate change—is hidden away in distant power plants.*

These monsters are not cuddly irritants, like the fluffy creatures in the 2001 movie *Monsters, Inc.* On the other hand, they need not be the world-ending biblical four horsemen of the apocalypse: pestilence, war, famine, and death. To see why not, we need Latour's contribution.

• • •

Bruno Latour is a French sociologist, anthropologist, and philosopher, despite which, he has written an essay of practical significance. He describes Frankenstein's monster and draws an analogy to technology and our attitude toward both. Mary Shelley was concerned with humankind—represented by the well-intentioned Dr. Frankenstein—messing around with Nature; hence the analogy. Latour's idea, which seems to have been well reported and discussed (if not so well accepted) is this. The tragedy of Frankenstein's monster is not the fact that it was created, but rather that it was abandoned. Had the monster not been abandoned and rejected when it was seen to be and do things

---

*Here is a mischievous image: we use air conditioners to keep out the high temperatures that result from global warming, which is a consequence of electrical power generation that we need to power our air conditioners.

that were unwanted and unwelcomed, then it might have been made acceptable by further tinkering and tweaking of the "prototype." By analogy, climatologists, conservationists, and environmentalists should not abandon technology because it has led to unintended consequences (see chapter 39) that have been bad for the planet. Instead, we should persist with technology, because it is perhaps the best bet we can make for getting out of the jam it has put us in. Thus we should nurture genetic modification and nuclear power as we would nurture our children—raise them to be useful members of society in their maturity, not reject them as unworthy of our parentage just because they have disappointed us.*

Latour's essay is called "Love Your Monsters," and it places its author firmly in the *technofix* camp. Technofix is one of four approaches proposed as the solution to the climate problems we face as we enter the Anthropocene—it and the other three are summarized, paraphrased, or lambasted in a later chapter (31). I am probably giving nothing away by declaring that, in my humble opinion, technofix is the only sensible way forward. However, some enthusiastic denizens in this camp (it seems to me) get breathlessly carried away with the technological possibilities, without considering their feasibility. Technology can only be the answer if it is implemented, and implementation requires more than just compatibility with the laws of physics. To implement a technology that can have effects on a planetary scale, we must as a species agree to do it—the people in one research lab or one nation are not enough. We must all pay for it, build it, deploy it, and accept its consequences. This is a big problem, which I hinted at in the introductory chapter. Getting enough people to agree to a large-scale project, and to cooperate effectively for long enough so that it is successfully seen through to completion, is a difficult and perhaps impossible task. Our collective intelligence and competence are less than those of individual people. Crowds are stupider than individuals that form them. This characteristic of humanity—seen over and over again in politics and in industry, and examined in chapter 40—takes the edge off any optimism for a technofix solution to our collective problems. I will explore the difficulties of large-scale cooperation in several later chapters.

*See Latour's essay in Nordhaus and Shellenberger (2011)—the other essays in this collection are also worth a read. See also Horgan (2011).

28 Scottish Philosophy and Nuclear Power

One Scottish philosopher—I think we can describe Adam Smith as such, for much of his work was theoretical and of that ilk—has influenced history mightily, ever since he penned his magnum opus *An Inquiry into the Nature and Causes of the Wealth of Nations* and so kick-started the modern discipline of economics. Of course, economics had been practiced for millennia before Smith ever graced the Scottish Enlightenment in the late eighteenth century—it has been practiced by everybody who ever lived. The gentleman who first traded a chicken for a spear practiced economy as did, for that matter, the lady who started the world's oldest profession. However, Smith studied the theory behind it—economics, not prostitution—and his followers and detractors (for his influence spreads well beyond the capitalist world he analyzed) erected what would be the global edifice of modern economic theory. One practical realization of current economic theory is the globalization of production and trade, as we have seen. Another, arguably, is the tedious repetition of financial crises that bother the stock markets and banks and investors and housebuyers of the world from time to time, reminding us all that nobody, but nobody, really understands economic theory. Oh, Adam Smith ably navigated the pond of eighteenth century capitalism and how it benefited the then-developed world, but he would have been cut adrift in the tumultuous ocean of twenty-first-century globalism.*

Scots of the Enlightenment had a great deal of influence on the modern world. David Hume was another such—perhaps less influential on the lives of modern humans, but certainly still a major player

---

*Smith also wrote *The Theory of Moral Sentiments*. I don't know what he would have thought of the moral sentiments behind the financial crisis of 2008, though surely people and the manner in which they engage with each other have not changed much in the last 250 years. Reaction by economists to the 2008 crisis shows that they don't understand it—or, more accurately, that there is nothing like a consensus among the experts. Compare, for example, the reactions of Professor Paul Krugman and Professor Niall Ferguson—they have diametrically opposite prescriptions for recovery (see chapter 34).

in the field of philosophy. An amiable character by all accounts, Hume it was who told us that human rational thought springs from our emotions.* We are, perhaps alone among life forms on Earth, uniquely capable of rational thought. We can abstract and reason about matters that do not involve our immediate needs for food, sex, or safety. Nevertheless (and here we arrive, in a very roundabout way I admit, to the reason why I include eighteenth-century Scots in an essay about nuclear power) humans are at base not rational. Every rational thought arises in, and is overwhelmed by, the same brain that evolved emotions—think of a salad that is drenched by dressing.

Thus, a mathematician may seek to be a leader in her field by the application of her rational brain, but why does she want to be a leader in her field? To gain the esteem of her peers, perhaps. A man works and thinks hard to enrich himself beyond his immediate needs. Why? To impress the ladies, perhaps. Neither aim is rational, though the means to achieve it are. We humans are not rational, as Hume said, and as economic theorists are now realizing. If ever the rationality and irrationality of humanity were on public display, it is in the arena of nuclear physics and nuclear power, respectively. And so we turn again to that neglected source of our future power.

• • •

We have seen why it is desirable for humanity to steer away from fossil fuels and toward renewable sources of energy. We have seen how this is attainable—to a degree, note—as a result of the technological innovations of recent, and surely of future, decades. Winds are the most promising renewable energy source in the short term. Wind power is renewable because wind energy is derived from solar electromagnetic energy (the sun's rays heat the Earth's surface differentially, leading to rising air and horizontal pressure gradients). The US Department of Energy is sufficiently encouraged by the prospects of wind power to suggest it might contribute as much as 20% of the national power requirements by 2030, as we saw in chapter **22**. Solar power is derived directly from the sun, as I hope my readers may have

---

*In a nation of ardent believers, Hume was an atheist, who died at peace with the world. His views on the human mind actually stressed the dominance of desire, rather than irrationality explicitly, over reason.

guessed from the name, but so far the technology eludes us for making this obvious source economical—though this situation is changing. Both wind and solar power are difficult to integrate into national power grids because they are irregular or unreliable (see chapter **26**)—the wind iconically so, and solar power between day and night.

Today's best source of renewable energy and power is provided by hydroelectric dams. These provide clean power that is economical, and can be controlled so as to meld nicely with the requirement of national electricity grids. The problem with hydro is that it is geographically variable: some countries, such as Canada and Norway, are blessed with rain and mountains, and so have the potential for ample hydropower; others, such as the Netherlands and Saudi Arabia, are not—there is tension between nations over water for reasons of power as well as irrigation. Thus, as we saw in chapter **22**, geography and geology conspire to limit hydro's share of global power sources to under 10%.

We can currently envision renewables as providing no more than about half of humankind's current electrical power needs. Invention, innovation, and technological breakthroughs (see chapter **27**) may cause the number of renewable megawatts to rise, but the power needs of humanity are also due to rise, as populations and the wealth of developing nations increase (see chapter **11**), so the fraction of generated power that will be met by renewables will not increase enough to meet our growing needs. If fossil fuel use drops, due to environmental/ climate concerns or due to the running down of fuel sources (oil, natural gas, or coal fields), and if renewables can't make up the shortfall, then from whence, you ask, are we to purloin our power? You guessed it, from inside the nuclei of uranium, thorium, and plutonium atoms.

• • •

Science is the application of rational thought, plus rigorous observations, to unravel the secrets of nature. The success of science is astonishing; as a field of human endeavor, it is surely the most successful. For example, note that quantum field theory (QFT) can make predictions about nature that are accurate to 11 decimal places. What other description of nature can claim such results? Indeed, what field of study outside science can quantify its success at all? My friend Dr. Michael Lowe—a physicist I shared an office with while at graduate school

35 years ago—expressed this thought very well. Look at the ancient Greeks, he said. Their philosophy, art, literature, and theology have been admired and well studied for millennia. A philosopher can say that his subject has progressed since the time of the Greeks—so, no doubt, could an artist, writer, or theologian. But how is an ordinary person on the street to know? I can see, for example, that Renaissance art is different from the art of classical Greece, and an expert art historian might be able to convince me that the former was influenced by the latter, and not the other way around, but I cannot see any obvious progress. Similarly for the other disciplines: differences but no obvious progress over the last 2,500 years. Yet with science, *anybody* can see the difference between that of ancient Greece and that of today.

Nuclear physics, as a discipline, is closely related to quantum field theory. For sure, its predictions are not as close to nature's truth as the best of QFT, but nuclear physics is a quantitative science that has made significant progress since atomic nuclei were first proposed over a century ago. The main applications of this theoretical and experimental scientific discipline have been nuclear weapons and nuclear power. These applications bring a technical research discipline into the limelight, placing the entire subject under the public gaze. Consequently, and because we are as a species largely irrational, the rigor of scientific analysis is replaced (not entirely, but to a great extent) by emotion. Political and economic considerations are mixed up and confused with science. "Nuclear weapons kill people and so are bad: science gave us nuclear weapons and so science is bad." Of course, the view just expressed is a simple-minded pastiche, yet I recall it from some radical fellow students back in the day. More commonly, a fear of nuclear weapons has led to a fear of nuclear power, and a suspicion (at the very least) of nuclear physics.

There have been accidents at nuclear power plants that have led to death and destruction. These have been due to errors made by human operators, or power plant designers, or acts of God, or perhaps, by the time this book goes to press—who knows?—by acts of sabotage. None have been due to faulty nuclear physics, but that is not the point, except to a defender of science. To a member of the public, all that matters is the safety and efficacy of the finished product and so physics, technology, design, management, and security all have to work. That they don't is human nature. As a result of the major disasters at Three Mile

Island, Chernobyl, and Fukushima Daiichi, serious radiation leakage has affected tens of thousands of people and will kill perhaps 4,000 (almost all from the Chernobyl disaster—the world's worst, by far). Billions of dollars of damage was done (cleanup, upgrades, replacements, screening, and treatment) and large areas of land are uninhabitable, and will be for decades. Three Mile Island and Chernobyl led to a cancellation of nuclear power plant building programs for a generation; Fukushima happened just as public confidence was returning (see chapter 25).

Airplanes drop out of the sky because of design faults, maintenance schedule cock-ups, terrorism, mentally unstable pilots, enemy action . . . you name it. According to the Geneva-based international Aircraft Crashes Record Office, 17,849 people have died in air crashes this century (including 9/11). Yet how many people call for a moratorium on flights? None, to the best of my knowledge. This response is natural, human, and irrational. Democratic governments act in such matters according to the wishes of their people, by and large, and so the irrational views of many individuals percolate up into irrational international actions. I have argued that much of this irrational response is due to the statistical nature of nuclear radiation. Our brains are not good at making statistical judgments, particularly when the odds are long, as many studies have shown—see the subtly titled chapter 29 for a telling example.

Aircraft fatalities can be established very accurately, but radiation fatalities cannot; this is the problem. The figure for Chernobyl is a best-estimate average from sober sources, but other sources, with their own agendas, claim the true figure is as low as 31 or as high as a million. Radiation from nuclear fallout can induce cancer deaths decades after the event, so many of the 4,000 will not have died yet, or even become sick—the estimate is based on statistical calculations. We might attribute to the disaster all cancer deaths in the Chernobyl area that exceed the average number before the accident occurred. But is this fair? How do we define "the Chernobyl area"? What if, as is very likely the case, the demography has changed in this area over the decades since the accident? You see the problem.

So people think like this: I don't understand the statistics, so to be on the safe side I will assume the worst-case scenario. This view is silly and wrong—we don't think like that when it comes to flying in a plane, or having an X-ray, or sunbathing—and this attitude is dangerous, as I will argue below.

• • •

Nuclear fuel is, for all intents and purposes, renewable—it's a finite resource with a natural supply running into thousands of years. Nuclear power on a normal day generates little toxic waste and contributes very little greenhouse gas to the atmosphere. The main environmental effect of nuclear power is that it heats up surrounding river or sea water a little (because many nuclear power plants use water to cool the generators).* This is a mild effect, compared with any other source of power—including hydro. Nuclear power plants and the associated fuel mines take up a small fraction of land, compared with other sources of power, such as solar arrays or wind farms (though wind farms can be placed offshore, and onshore wind farms can be placed in potato fields, for example, so that they can be made to hog little space). Nuclear power is a quite mature technology, and the production cost per megawatt-hour of energy produced is the lowest of all major sources, if the power plant is required to operate at high capacity (above 70% of peak capacity—for lower-capacity operation, combined-cycle gas turbines are the most cost effective). Nuclear power plants can be built on any stable ground, and they are easy to integrate into large electric utility grids. It's a no-brainer that we should be using nuclear power more than we are, and should be building more nuclear plants for a fossil fuel–free future. The French obtain over three-quarters of their power from nuclear reactors, and suffer far, far more deaths from drinking wine than from nuclear accidents. Should we ban wine?

By not using our knowledge of nuclear fission, and our accumulating knowledge of how to harness it to generate power for us, we are committing ourselves to burn more fossil fuels than we would do otherwise. We have seen in earlier chapters how these fossil fuels commit us to future climate change. Our not using nuclear power is more dangerous for future generations than our using it now. Put another way:

---

*Including Fukushima—this is why the reactors were built near the coastline. You might ask why nuclear reactors were built along a coastline that is known to be at risk from earthquakes and tsunamis.

we will suffer less from the bad side effects of nuclear power than our descendants will if we don't use it. We have already borrowed a lot from them, in terms of climate and (as we see in chapter 30) money. By not employing nuclear reactors today, we are borrowing even more from their wellbeing.*

29 You Suck at Statistics

We have all heard the joke about the statistician who drowned trying to cross a river of 3 feet average depth. Well, he must have been a pretty bad statistician as well as a bad swimmer. A competent statistician would have asked for the standard deviation of the depth, and maybe for some higher-order statistics, before dipping his toes in the water. This section will serve to show you that most people would drown in that river.

• • •

To see that your intuition about statistics leads you wildly astray, especially when the odds are long, consider the following medical example. You are concerned that you might be suffering from a dreadful ailment, the Screaming Hab-Dabs, which randomly affects one person in a thousand around the world. You visit your doctor, who applies the standard 3H-D test. This test has a false-positive rate of 5% (meaning that 5% of the people testing positive do not have the ailment) and a zero false-negative rate (meaning that, if the result of the test is negative, you certainly do not have the ailment). For you, the test result is positive. What is the probability that you have the Screaming Hab-Dabs?

If you answered "95%" then you are among the vast majority of patients, doctors, and even mathematicians around the world. Piffle and balderdash, a statistician would say. The actual probability is 2%. Quite a difference, eh? Here is the proof. Select 1,000 people randomly from

*For an article about the dangers posed by our fear of radiation, see Rockwell (2011). For economic betrayal of the next generation, see, for example, Barr and Malik (2016).

the population. We know that, on average, 1 of them will have SH-D. But 51 of them will generate positive test results. So 1 out of 51 has the ailment. Voilà.*

• • •

We now pause to take stock. In the next chapters I summarize the story so far—where we are, how we got here—and then see where we are likely to be heading in the short term.

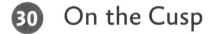

30 On the Cusp

Humankind is altering the planet and will soon enter the Anthropocene Age, if we have not already—thus formalizing the notion. The changes that we have made to the Earth are due mostly to our technology, which has led to environmental degradation and probably to the observed recent mass extinctions, and is leading to significant climate changes. Technological advances have continued apace for 250 years and are changing society—we are currently creating and experiencing the third industrial revolution. As a direct result of applying science to food production, our total population has increased beyond the level that can be sustained without such science. Human numbers will continue to increase and will probably level off in the second half of the century at around 9 to 11 billion. Global atmospheric temperature will also level off, but when and where is uncertain. A 2° Centigrade rise (at the least) is now probably inevitable, whatever we do to mitigate atmospheric carbon dioxide levels,† due to the amount of CO_2 already present in the air; a key question is, how high will average atmospheric temperature go? If it increases more than 4 degrees, global climate models suggest, there may be runaway and unavoidable

*Would it be confusing and unhelpful if I reveal to you that the main symptom of the Screaming Hab-Dabs is a complete loss of understanding of statistical concepts? Probably. So let's say that the main symptom is a hacking cough, such as that of Professor Albedo.

†One climate expert, Kevin Trenberth of the National Center for Atmospheric Research in Boulder, CO, believes that we will reach a 2°C increase by 2060. See Greenfieldboyce (2015).

changes that lead to a new climate that is hostile to human life. The drop and subsequent rise in ozone levels show that humans can fix as well as cause changes in atmospheric constituents, and so offer hope for the future, though carbon dioxide levels will prove to be a much harder nut to crack than CFC levels.

To address global warming due to anthropogenic CO_2 generation, humanity is beginning to change energy sources from fossil fuels to a mixture of renewable, or sustainable, technologies: principally so-lar, wind, and nuclear power. Delivery and security of a steady and reliable electricity supply will require major upgrades to current electricity grids toward so-called smart grids. If in the future, technology is to work for us sufficiently to solve the climate and environmental problems we have created for ourselves, we must embrace our technological monsters and nurture them.

Such is the state of the world as we poise on the brink of the Anthropocene.

• • •

What of the future? There are two types of crystal ball gazing to be indulged: short term and long term. Gazing into the near future is relatively easy—it consists of linear extrapolations, of observing current conditions and then extending current trends. Such predictions generally do not require much expertise beyond statistical analysis, except that we need experts to tell us what the state of the world is at the present time and how things are changing today. Long-term predictions are much less certain; they involve a detailed knowledge of the underlying dynamics of the human world, and such knowledge we do not possess—our long-term "predictions" are really little more than speculation. Nevertheless, I will speculate about what life will be like a century hence, in chapter 42—take such ruminations with a large pinch of salt. Here I will make the easier and much more likely short-term predictions, the linear extrapolation of current trends into the coming decades. The confidence that we can place in such predictions falls away from near certainty—no salt required—for the 2020s as we look further and further ahead, to the second half of the present century.

Population will increase to something over nine billion and will then probably level off, as we have seen (chapter 11). This is the percentage estimate—there are outliers that may occur, such as a catastrophic fall in population as food runs out (Malthus) or as climate change turns the Earth against us (Lovelock—see chapter 31), but here I will stick with the most probable scenarios. The change in population will be accompanied by a change in demographics—humanity will age. Today some 11% of people are aged 60 or older; this fraction will increase to around 25% by 2050. Certainly in developed countries, and to a lesser extent elsewhere, people will enjoy extended periods of wellness and a shortened period of decline and disability in old age. Increasingly sophisticated biomedical devices will revolutionize the diagnosis and treatment of certain medical conditions such as cancer and Alzheimer's. Dementia will become relatively more common in low- and middle-income countries than in wealthy countries (increasing from about the same numbers today to perhaps twice as many by 2050). Just as iron lung machines have been made obsolete by the polio vaccine, so many of today's device-based and instrument-based therapies will be replaced by molecular- and cell-based therapies. Therapeutic cloning will increase, as will genetic engineering to treat or eliminate diseases and aging. It will be technically possible to preselect genetic traits in newborn children, though to what degree such eugenic steps will be socially acceptable is harder to predict. Obesity will continue to increase, especially among the young (see chapter 13). Medical health insurance and treatment costs will increase a lot, perhaps breaking numerous national social medicine programs; in the United States, Medicare and Medicaid costs are projected to more than double, to 12.4% of GDP by 2050.

Most human beings will be city dwellers by 2050. Their social world will be less benign than ours, with reduced Social Security benefits and increased extremes of wealth, poverty, education, and health compared to today. (I have not yet discussed rising inequality—see chapter 32.) In the United States, race issues will be less virulent and manifest than today, but class issues will come to the fore, thanks to rising social inequality. Whites, excluding Hispanics, will be a minority (at 47%) of the American population; projected figures are 29% Hispanic, 13% Black, and 9% Asian. The global use of illicit drugs will increase in line with the population, and most of this increase will be among the new city dwellers. The internet will kill off traditional news media, replacing them with smaller, niche news organizations.

We will all be encouraged, perhaps legislated, to recycle and reduce our carbon footprints.*

Food production will have to increase 70% by 2050 to provide the nourishment required by a burgeoning population. About half of this increase will come from increasing crop yields (projected to be 0.8% per year—lower than past values, as discussed in chapter 12). The rest must come from increased cropping intensity (more harvests per acre per year) and increased acreage devoted to food production. The expansion of agricultural land will be uneven: it will increase globally by about 5%, but will be higher than this in developing countries and will actually decrease in the developed world. Boosting irrigation in developing countries, coupled with the effects of climate change in many of these regions (see below), will place even greater stress on water resources. Agricultural trade will increase, and the pattern of food production will change—for example, Latin America is expected to become self-sufficient in cereals, which it currently is not. Chronic undernourishment is expected to decline globally from 12% (2012 figure) to around 5% in 2050. Of course, a lot of ducks have to get into a row for all this extra food production to be realized, and history warns us that the growth will not be smooth or equitable.

Forced mass migrations of people historically occur due to political and social unrest. Today we see examples of these in the distressing images of migrants from Syria, Iraq, and Afghanistan inundating Europe. Economic migrants are also significant in numbers, from Central America to the United States, and from Central Africa to Europe, for example. Both these causes of migration will continue—it was ever thus—but in addition, the river of migrating humanity will be swollen by people who are moving because the climate is changing.†
Rising sea levels will displace people, as will increasing temperatures that will render infertile some arable land. Fresh water will become scarce in some parts of the world, and this will lead to social unrest.

*For future illicit drug use projection, see Travis (2012). For news media changes, see Master (2009).
†It is possible that the Syrian migrant numbers are partly due to climate warming effects; there was an extreme drought in the country from 2006 to 2011. See Baker (2015).

"Water wars are coming" (Mogil, 2007). In a recent article in the prestigious American magazine *Foreign Affairs*, we read the following: "The United Nations High Commissioner for Refugees (UNHCR) estimates that by the end of 2014, nearly 60 million [people] had been forcibly displaced owing to persecution, conflict, and human rights violations—the highest level on record—and of these, it classified nearly 20 million as 'refugees.' To these huge numbers may plausibly be added tens or even hundreds of millions more who would likely be attracted by any available option to migrate away from conditions of deep poverty, starvation, or environmental disaster" (Teitelbaum, 2015). The decades to come will be unsettled. We have borrowed from future generations, the bill is overdue, and they will pay it.*

● ● ●

The societal predictions provided above are hardly controversial, and will not be disputed by many knowledgeable people. These predictions may be modified by two key factors that will pervade the consciousness of our grandchildren and their grandchildren and guide their actions and attitudes more than they have guided ours: energy and climate. Predictions concerning the latter are in some ways no more plausible than those for society, because of the complex dynamics of climatology (see chapters 14 through 18), so their consequences for the lives of people are less certain. For example, it is just about possible that human intervention will make climate change so gradual as to have little effect on the lives of most people, or that it will be so rapid and severe that it will destroy human civilization, whatever we do to try to stop it. I do not think that either of these extremes is the most likely outcome. However, declaring that I tread a middle path is hardly enlightening to you, given the extremes. Most (though not all—see chapter 31) intelligent and informed opinions hold to a middle path. My view is less optimistic than those of many commentators, for

*See Alexandratos and Bruinsma (2012), Griffith and Grodzinsky (2001), Master (2009), McCarthy (2015), Olshansky (2009), and Teitelbaum (2015). The quote about water wars is from H. Michael Mogil (Mogil 2007), a meteorology consultant and writer. The figures for chronic undernourishment are from the UN Food and Agriculture Organization (http://www.fao.org/news/story/en/item/161819/icode/).

reasons that I will present and defend in the remainder of this book. In brief, it seems to me that our response to the energy crisis (I think we can call it such) and to climate change will be slow and incomplete, and that such foot dragging—unavoidable because of the world's prevailing culture of selfishness and hierarchy—will make a bad situation worse.

First, let us consider energy. There will be a sustained move away from fossil fuels and toward renewable energy sources (see chapters 22 through 25). Fossil fuel use will decline from 2020 but will still be significant; such fuels will have been made more acceptable by CCS and other filtering techniques that reduce emissions, but by mid-century they will still be adding carbon dioxide and particulate matter to our atmosphere. We will persist with fossil fuels because of our increasing need for energy, because of their low cost, and because of the powerful fossil fuel lobby.

Renewable energy sources, mostly solar and wind, will constitute perhaps half of all our growing energy needs.* Hydro will continue to play a useful but minor role in global power generation. Solar power will take up a lot of land area, but not land that is needed for agriculture. Storage of solar power will require efficient batteries; these and the disposal of old solar panels will create a significant but not overwhelming environmental problem. Wind power will dominate the renewable sector in the 2020s before solar power costs fall, around 2030, to the extent that solar takes over. This is the only way that a technology can come to the fore—by being cheaper than its competition, not by legislation or subsidies, though these incentives help push things along. Wind power will also take up a lot of space, and wind turbines will perhaps be a commonplace eyesore, but turbines can be placed offshore or in crop fields or pastures, so they need not take land away from food production or housing. Nuclear power, of necessity, will play a major role for the remainder of the century, increasingly so as fossil fuel use falls away in the second half of the century. Dilatory

*Bloomberg, the financial services and mass media company, has written a report on future energy and power that predicts an expenditure of more than $12 trillion over the next 25 years on renewable power development. The world's power generation capacity will double by 2040, they claim. See Kahn (2015).

action across the globe in commissioning new nuclear plants (caused by concern about, and misunderstanding of, the threats to safety posed by this technology) will extend the lifetime of remaining fossil fuel plants, to the detriment of our climate and environment.

• • •

Which brings us to climate change predictions. The extension of fossil fuel power plants means greenhouse gases will be dumped into the atmosphere for decades to come—the levels will increase over 50% by 2050, so that atmospheric CO_2 concentration will exceed 600 ppm. (GHG emissions from power plants are likely to peak around 2030, but for decades after will still be above present levels.) We have seen (chapter 16) that the global warming effects of GHGs, particularly CO_2, are long lasting, so the planet will continue to warm for some time, a warming that will almost certainly take us beyond the 2°C guardrail (see chapter 17). This is a prediction we can take to the bank—hmm, perhaps an unfortunate metaphor these days, given the recent financial meltdown and bank bailouts. I mean that we can be very confident in this prediction, but unfortunately we cannot be so confident about most long-term climate predictions, despite the increasing sophistication of climatologists' computer models (GCMs). Why? Because physics is not the only influence on climate change. We have seen that geology can throw off climate model predictions via volcanic activity; more importantly here, economics will have an increasingly important influence in the Anthropocene age. (Volcanic activity cannot be predicted decades in advance, and economic activity can barely be predicted from one month to the next.) I will argue this case here and in chapters 33 and 34.

We are at a cusp and on the horns of a dilemma. The cusp is the dawning of the Anthropocene age: our future depends very much on what we do now. The dilemma is deciding what action to take, to avoid damaging (at best) or catastrophic (at worst) climate change and—equally important—getting everyone on board. Former US secretary of state John Kerry has said this about climate change: "Unless we act dramatically and quickly, science tells us our climate and our way of life are literally in jeopardy . . . The costs of inaction are catastrophic" (McGrath, 2014). Indeed, the science is pointing that way. This global problem requires a global solution, with all nations contributing to a common effort to avoid Kerry's catastrophe and here, alas, is where we will fall down, in my opinion. History shows few examples of lasting cooperation between nations, especially when their interests diverge, and shows many examples of irrational or ill-thought-out actions that are self-defeating.

In later chapters I analyze our very human inability to act in concert—see **37, 39,** and **40.** Accepting for now that quick and decisive global action is needed, and that it will not happen (or more precisely, it will happen, but too little, too late), how will our environment change in the decades and centuries to come? The longer-term predictions are really speculations, recall, because our knowledge of climate change is imperfect and, more importantly, because *long-term* climate evolution depends in part on nonscientific factors such as economics, which is unpredictable. So we don't really know what climate effects are coming at us, some way down the road, let alone what to do about them. *Short-term* climate evolution is clear enough from the GCM models and is more believable, because there is less time for unpredictable effects to screw up predictions; for those of us in developed countries at temperate latitudes, we can expect:

- Increased instances of flooding, and rising sea levels
- Increased periods and severity of drought
- Changing fish catches, as fish populations move due to escalating sea temperatures
- Changing crop yields, due to alterations in atmospheric temperatures
- Increased numbers and severity of wildfires
- Increased instances of extreme weather, such as hurricanes and tornadoes.

At different latitudes, the effects will be worse, as we will see later. Over the long term, the climatologists tell us, they will also worsen at temperate latitudes.

Domestically, we can address potential flooding by building or extending sea walls and levees. Drought effects can be mitigated by extending irrigation and by securing water resources—the latter is not easy to do and is next to impossible, due to global warming, if your water comes from snowpack rather than rainfall.* The fishing industry will be disrupted, as will farming. We must prepare for disaster relief when hurricanes or tornadoes strike. We must site the new and extensive wind and solar power plants (and new nuclear plants) out of the way of hurricane and tornado hot spots.

These consequences of climate change are internal to each nation, and each will deal with the problem in its own way. The real problem is international cooperation to deal with global causes as well as effects, such as polar ice melting, and increased atmospheric temperatures (perhaps 3°C–6°C by 2100) caused by carbon dioxide levels. If we all must reduce fossil fuel burning, what if someone cheats? And someone—perhaps everyone—will cheat (see chapter 37).†

 **31 Four Fixes**

What are the fixes available to us, to guide our wayward species into a secure Anthropocene age? Here are four categories that have been much discussed, four widely different approaches that reflect greatly different points of view on the nature of our present and future. Actually, only three of them represent fixes—the fourth viewpoint is that there are no fixes.

• • •

Business as Usual is the label attached to the ostrich-headed approach that nothing is wrong with what we are doing now, so we should simply proceed with more of the same. This view maintains that changes in climate, if they exist at all, are due to natural fluctuations that have nothing to do with human industry, so any attempts to limit coal burning or to tax carbon emissions are a drag on industry that will only

*Bolivia gets most of its water from snowpack or melting glaciers; a state of emergency was declared in November 2016 due to water shortage. See Cohen and Ramos (2016).
†See McGrath (2015), Marchal (2011), and Stix (2012).

pander to a left-leaning intelligentsia, while threatening jobs and profits. Not surprisingly, the Business as Usual (BaU) scenario adopted by the IPCC in their climate models leads to the greatest increases in GHGs of all their scenarios.* Equally unsurprisingly, the BaU viewpoint is not common among climate researchers. BaU tends to be adopted by conservatives, some of them blue-collar workers, who see themselves as potential losers if the current climate debate leads to significant actions that impose emissions restrictions. Stephen Harper, the recently deposed Canadian prime minister, under whose leadership the Alberta tar sands industry has thrived, belongs to the skeptical BaU school of thought, along with many members of the Republican Party in the United States; President Trump expressed views along these lines during his campaign.

The last paragraph is as fair as I can be to these boneheads: in my opinion, their view is dangerous and courts disaster. The science (which many of them view with some contempt) is clear on this point— or, I should say more accurately, as clear as it can be given the impossibility of making long-term predictions (see chapter 30). The influential BaU group exploits the uncertainties to which statistical data give rise, so as to direct policy to their advantage.

● ● ●

The *Love, Peace, and Granola* label attaches to a quite different set of people.† The viewpoint adopted here is that if we all act in an environmentally responsible manner as individuals, and vote Green, we have done our bit and it will turn the planet around. Smart cars and hybrids sell to this group (even if the hybrids have to be hooked up to a coal-

---

*The BaU scenario leads to an average increase in global temperatures of 0.3°C per decade for the remainder of the twenty-first century, according to the IPCC, which would put us north of a 5°C rise by 2100. The IPCC adopts different scenarios within their climate models, in an attempt to account for the unpredictable effects of human development. The definition of BaU differs between climate models. See Teng and Xu (2012).

†My label for this group comes from a quote from David Crosby, of Crosby, Stills, Nash and Young, who once said of their songs, perhaps cynically: "That love, peace and granola shit went over real big, didn't it?" See Zimmer (2004). Of course, not all optimists are of this ilk; for the soberly optimistic view of a seasoned political journalist, see Friedman (2008).

powered grid). Local produce will guide these well-meaning people: count the miles that the zucchini took to come to market. Recycle assiduously, buy organic eggs (or—better—house your own chickens), advocate low-tech solutions to infrastructure problems in Africa, and aspire to live off the grid.

The trouble is that, while such a viewpoint may give a warm fuzzy feeling to the affluent westerners who espouse them, it is incomprehensible to many people in the developing world, where much of the environmental damage is being done (see chapter 5). In the developed world, too, by far the greatest amount of GHG emissions and pollution comes from industry and the power generation sector (see chapter 26), not from consumers. The Love, Peace, and Granola (LPG) fix won't—it's far too late for such minor titivation to make enough of a difference.

• • •

Technofix. I will devote more ink to this fix than to the others because it has a chance of working. There are two approaches to reducing or reversing the effects of anthropogenic climate change. The top-down method is authoritarian, and requires the imposition or restriction of actions upon people and institutions: limiting electrical power, heating, car use; controlling land use and imposing carbon taxes, etc. The bottom-up approach is technological, and is more palatable to citizens of liberal democracies, who often bridle at legal impositions. The technological fixes proposed for climate change are of two types: short-term "band-aid" fixes and longer-term, more sustainable, and perhaps acceptable innovations.

Even among technophiles, the band-aid technofixes meet with a quite varied response, from eager enthusiasm to profound concern. They are collectively labeled "geoengineering" and consist of a raft of ideas for reversing climate change via feasible, but as yet largely untested, technology applied on a planetary scale. One common suggestion is to increase the atmospheric albedo (reflectance of sunlight) by seeding it with seawater droplets. This action will reduce the solar energy that is absorbed and thereby impede the driver of global warming. Surface albedo can be reduced by planting pale crops and painting rooftops in pale colors. The ocean surfaces can be made more reflective, it seems, by blowing microbubbles of air into them from generators installed on commercial shipping. Carbon dioxide in the atmosphere can be removed by planting trees, both real and artificial. If the oceans are seeded with iron, phytoplankton will bloom and remove atmospheric carbon dioxide. A natural mineral—olivine—can be pulverized into a powder and spread from airplanes; seemingly, it will bind atmospheric CO_2 to its surface and then fall out as a solid precipitate.

These actions are more or less feasible, but will cost a lot of money if applied worldwide, and will take a great deal of energy to get up and running. The aerosol idea—spraying seawater high up in the troposphere or stratosphere—will certainly work, and quickly; we know this from the prompt cooling effect that powerful volcanic eruptions (such as that of Mount Pinatubo, in 1991) have had on global temperature. There are two serious problems with all these geoengineering fixes, however. First, we are tickling the dragon's tail again. We are fiddling with the natural order of things—a natural order that we do not understand enough to alter with impunity. Opponents of geoengineering foresee the same natural backlash that resulted when humans introduced new species into an environment—cane toads into Australia, for example, to reduce pests (see chapter 39); the results were disastrous. It is easier to release aerosols than to recapture them; we might not be able to control or undo what we set in motion, if it proves to have been a mistake. The second problem has more to do with human nature than planetary nature, and worries me more. There is a moral hazard associated with geoengineering; if people believe that geoengineering will work, they will feel less inclined to restrict carbon emissions—we can continue polluting beyond the red line because we have the means to step back from the red line later. It is like the apocryphal Volvo drivers; these cars have a good safety reputation, so some of their drivers feel OK about driving more recklessly.

The long-term, safer technofixes revolve around energy sources and storage. If we can design power plants that provide us with the energy we are going to need in the centuries to come, that does not contribute to global warming or pollute the environment in other ways, we will have gone a long way toward eliminating the source of our climate woes. Fusion power is the great hope here, with fourth-generation nuclear fission power plants as the intermediary that will see us through to the decade when fusion becomes technologically and economically viable. (The label "fourth generation" refers to fission reactors that can burn spent fuel safely and almost completely, leaving only 2% of the radioactive waste that is left by a present-day fission reactor, and therefore much reducing the waste disposal problem.) Another long-term energy technofix is the design and commercial implementation of large-scale electrical energy storage devices. At the present time, the devices we use to store electrical energy (batteries, capacitors, etc.) are inefficient and quite expensive.

Another technofix, less headline grabbing but perhaps of some considerable importance, is the notion of trash vaporization. Garbage burned at very high temperatures is reduced to its elemental constituents, many of which (metals, particularly) can be recycled. Thus two birds can be killed with one stone: we alleviate one environmental problem (energy sources) by burning another (accumulating trash and pollution). Trash vaporization technology exists—so the idea has been proven physically—but, as always, the real difficulties emerge when trying to take an idea that works in the lab out into the commercial world. It will take decades to make trash vaporizers cheap enough so that it becomes worthwhile to make lots of them—enough to make a difference.

All this clean energy that (technophiles believe) will be on tap sometime in the fuzzy future will be used for purposes other than direct electrical power generation. It will also be used to produce fuels such as hydrogen, which have a high energy density and burn cleanly. It will be used to make steel—replacing coke, a fossil fuel derived from coal, which reduces iron ore to elemental iron, from which steel (an iron alloy) is derived. Large amounts of energy will also be needed in the future to alleviate the coming water shortages via desalination. There are lots of ideas on the drawing board; some have progressed to the blueprint stage and beyond. Whether or not these technofixes are realized depends on whether or not we can love our monsters (see chapter 27) enough to nurture them. We must also choose wisely which of them to nurture, and how we use them.*

*For a range of views on technofixes in general, and geoengineering in particular, see Achenbach (2015), Gillies (2014), Goldenberg (2015), Hall (2015), Helm (2012), Huesemann and Huesemann (2011), Pielke (2011), and Shukman (2014). Also worth a look as a useful summary is the online lecture "Techno-fixes for climate change"

•   •   •

We're Doomed. Sometimes spelled with an "f" (Zolfagharifard, 2014), this label is the nonfix—the viewpoint that says we are heading toward environmental catastrophe, whatever we try to do now. Some of these Doom 'n' Gloomers add a moralistic addendum: we deserve it, because of what we have done in the past. We can easily dismiss such hand-wringing self-abuse, which seems to be rather hard on the future generations of people who will suffer the consequences without having played any part in the causes. However, we cannot so easily dismiss the predictions, for two reasons: because of their magnitude, and because several of the people who make them are expert ecologists and climatologists who should know their stuff.

The pessimism of this view arises because of the latency inherent in climate physics: it takes a long time to build up a head of steam, just as it takes a long time to turn an ocean liner around. So the "doomed" idea is that the amount of GHGs that we have added to the atmosphere is already a fatal dose—the snake venom has been injected into our bloodstream but has not yet had sufficient time to wreak its havoc. That time will come and is now inevitable. The climate will change so dramatically that the human population will crash due to climate effects directly (drought, lost arable land, water shortage, famine) or indirectly (war—perhaps nuclear—due to population movements or competition for limited resources).*

Some of the worst scenarios are envisioned by authors who are journalists, and who perhaps have an eye on a bestselling read rather than

---

by Barry Brook, at https://www.youtube.com/watch?v=EZShFq-36MI. The moral hazard of geoengineering is well brought out by D. Keith in his TED talk "A critical look at geoengineering against climate change." http://www.ted.com/talks/david_keith_s_surprising_ideas_on_climate_change.

*The population will crash perhaps 80%, according to James Lovelock. That is, the population will rise and then collapse until it reaches something under two billion. Lovelock has been a prominent environmentalist for decades, and is best known for proposing the Gaia hypothesis. (Perhaps surprisingly, he has also been a long-standing advocate of nuclear power.) He sees a very bleak future for humanity, and suggests that we enjoy the next twenty years because we won't have much fun thereafter. See Aitkenhead (2008). Famine results from failure to develop new crops or from harvest failures of these crops—the latter made more likely by global warming. See McGrath (2015).

on heartfelt convictions. Gwynne Dyer's book falls into this category—he is an excellent speaker, and his book on future climate wars reads like a thriller. On the other hand, James Hansen is a longtime, well-respected climatologist (his testimony on climate change to a congressional committee raised awareness of the subject, in the late 1980s) and is an undoubted expert who feels—like Lovelock—that it is simply too late. It's too late to bother with remediation—ending or reversing our contribution to atmospheric greenhouse gases, or recycling or living green. Such activities amount to rearranging deck chairs on the Titanic, they argue—we have borrowed too much from the future, and soon it will be payback time.*

③② Über Alles

"The argument today has moved on—to the growing inequality that is a side-effect of new technology and globalisation; to the nature of employment, pensions and benefits in an Uberising labour market of self-employed workers." Thus opines the *Economist* in a recent editorial (September 19, 2015). What is this Uberized economy, and why does it lead to inequality? Is it a significant factor as we move into the Anthropocene, or is it a fleeting market trend that will soon be relegated to a footnote in textbooks of economic history?

Known euphemistically as the "sharing economy," this new arrow in the global economy quiver is a means by which individuals can monetize their excess or underused resources. The idea is that privately owned goods are shared or rented via peer-to-peer marketplaces, usually on the internet. You will know a lot about this way of generating income if you already rent out an unused room in your home via Airbnb, or rent out your services as a cab driver—the cab being your own vehicle—via Uber or one of its many imitators. More likely, you have heard of these two most prominent flagships of the sharing economy

---

*For some thought-provoking premonitions of impending disaster, see Aitkenhead (2008), Chomsky and Polk (2013), Dyer (2008), Hansen (2010), Scranton (2015), and Stix (2012). A sober assessment of the danger of inadvertent war between Putin's Russia and Trump's America is given by Stout (2016).

because you have used them—as a replacement for expensive hotels in vacation cities, or expensive regular cab services in any major city. Or perhaps, you have heard of them through the controversy and backlash they generate.

The sharing economy is also known as the "collaborative economy," the "on-demand economy," and the "gig economy," which gets a little closer to what it is about. Employment is distributed: a worker may never meet a coworker who signed up with the same internet facilitator. Uber's face to the marketplace takes the form of a smartphone app, which permits the public to dial the nearest Uber taxi for a local trip. Uber connects customer and driver, and takes a cut of the fee. Same for Airbnb and the many other services. You can rent out your driveway via Parking Panda, provide a canine version of Airbnb (dog-kenneling service) via DogVacay, rent out unused tools in your garage via Simplist, outsource household errands to willing or cash-strapped neighbors via TaskRabbit, or turn your cooking skills and kitchen to profit by making food and selling it via Krrb. You can swap stuff you no longer need for stuff that other people no longer need on Swap.com, or swap it for cash on eBay or craigslist.

No doubt some of these start-ups, and many others, will have gone bust or changed their name by the time this book goes to print. Equally certain is that more will have arisen. These digital clearinghouses are growing at a rate in excess of 25% per year, and in 2014 their collective revenue was estimated at $15 billion, and rising. Many began in the aftermath of the global financial crisis of 2008, and indeed they now provide much-needed income for many victims of that crisis. Your car is a classic example of an underused asset—it sits on your driveway 92% of the time; why not use it for your second job as an Uber driver, or rent it out via another sharing company? Or, if it is an electric car, perhaps you could generate some revenue by selling electricity to your local utility. Maybe you lost your regular job and can't get another, so now you have several small streams of income by providing many of the services described here. You rent out unused land for others to grow food, or rent out your boat—you don't have to invest anything that you don't already have; you don't need to learn new skills—you simply employ skills and resources that are already yours. The Uber economy is more than simply freelancing, like the handyman who advertises himself to neighbors through the pages of his local newspaper's classified ads. It certainly reflects previous trends—freelancing, contracting, temping, outsourcing—but is digitized and expanded to everywhere there is a smartphone, which is everywhere. A smartphone app permits sharers to transact their business anywhere, anytime. It functions in real time. It facilitates the flow of money—the business transaction. Uber could not possibly work via newspaper ads.

The facilitator—the company behind the smartphone apps—may itself have begun life as a product of the same kind of distributed exchange that they exploit, via the internet. Crowdfunding is an alternative source of financing a start-up, usually from many small contributors who get together through the internet. Some $5 billion were raised in 2013 by this method. Crowdfunding works where many regular banks wouldn't; banks may be unwilling to gamble, because of the hits or bad press they received after the recent financial crisis, or because they would be asked for a big loan rather than a small one.

Workers in the sharing economy have little power—this is where the inequality creeps in or, rather, charges through the front door. The many, many jobs that have been created by the sharing economy are part time, or temporary, or second or third jobs, with low income, no unions, no pensions, and few rights (and nobody to inform the workers of their rights). These aspects of the sharing economy help shrink the middle classes, who leak out at both ends—some becoming successful entrepreneurs, many more living hand to mouth as temporary, part-time Uberemployees. Thus inequality increases. The Uber paradigm is ahead of market regulators and lawmakers. Regular, traditional services fight back. Taxi companies hate Uber; hotels hate Airbnb; both fight back to restrict the newcomers, with regulations or taxes.

Many commentators applaud the new smartphone economic paradigm, while others bemoan it, or just declare it inevitable. How will the new companies—digitized cottage industries, yet with global reach—fare in the decades to come? The small companies will come and go; the survivors may or may not stay small. They will not always coexist alongside regular companies; rather, they may alloy with them. Just as airlines developed low-cost versions of themselves, just as the multinational brewing companies bought up-and-coming microbreweries, so, for example, the large car rental companies are buying into new Uberesque start-ups that share cars between individuals. Here is a possible peek into the future for a successful Ubereconomy company: Airbnb recently agreed to cover the expenses of a homeowner whose house was trashed by an Airbnb customer, and amended their rules of operation to provide a $50,000 guarantee to all their homeowners. This guarantee was later raised to $1 million, backed by Lloyd's insurance—Airbnb is doing business with the mainstream.

The mergers between new and old companies, the evolution into a more traditional mode of operation, will not be total, because the marketplace has been changed—for the foreseeable future and probably permanently—by smartphones. We call them "phones," but this is merely a nod to history; they are mobile computers and information exchangers, here adapted to a new type of business. So, it seems to me and many others, the Ubereconomy is here to stay in one form or another, and therefore societies that are free to adopt it (how can any society with smartphones prevent it?) will become less equal. Employees will be more anxious than in the past, with a less certain future in a more volatile market. How many people of your parents' generation worked the same job (for better or worse), from the time of leaving school to retirement? How many of your school friends have had only one job? Workers will be more mobile as well as more anxious in the Ubereconomy.*

• • •

The human world is changing, and we have changed it. There will be more social unrest and economic upheaval due to technology; bad climate changes are in the pipeline, due to technology and industry. Yet technology is our best hope for mitigating the worst effects of climate change, even as it upsets (for better and worse) the social order of the past.

How is the world poised for the type of collective action that will be necessary if we are to enable a technological fix for our climate problems? History and human nature are against us here. First we look at one big monkey

*Of course, much has been written about the Uberized economy—about what it is, what it means, and where it is heading. See Broderick (2014), the *Economist* (2013a), Freeman (2015), Geron (2013), Pittis (2014), Scheiber (2015), Spence (2015), Thompson (2015), Walsh (2011), and Younglai (2015).

wrench in the works—the hard truth that NOBODY understands global economics—and then consider our inability to act in concert for the common good (which does not augur well for a climate technofix). I will argue that these two factors are related.

33 Sherlock Holmes and the Anthropocene Deduction

One of the world's truly great fictional heroes, Sherlock Holmes has been popular across the world for a century. His fictional exploits, reported by his trusty biographer and friend Dr. Watson, have been in print during all this time. Some of his millions of devoted followers consider him to be a real historical person, who is still alive (after all, no obituary has appeared in the *Times*...) and who was initially represented by a literary agent—not creator—called Arthur Conan Doyle. This view can only have been encouraged by the surge in Holmesiana in recent years—plays, major movies, TV series—and yet Holmes as a living, breathing, crime-fighting human being is nothing new. The great consulting detective has been perceived as corporeal and approachable, despite his reputation for waspishness and aversion to publicity, judging by the many letters received at 221B Baker Street, London. For many years this address was an office of the Abbey National Building Society, which employed a full-time secretary to answer Holmes's mail.

The great man would time and again lament the lack of appreciation by lesser humans of his superior deductive powers. Thus, in "The Dancing Men:" "You see, my dear Watson... it is not really difficult to construct a series of inferences, each dependent upon its predecessor, and each simple in itself. If, after doing so, one simply knocks out all the central inferences and presents one's audience with the starting point and the conclusion, one may produce a startling, though possibly a meretricious, effect." This notion is a central theme of all the Sherlock Holmes short stories, novels, plays, and films. In "The Dancing Men," Holmes deduced Watson's thoughts while both sat in silence, and then broke in on them midstream, giving the effect of mind reading. When the chain of reasoning was explained, Watson nettled Holmes by saying how absurdly simple it seemed, once explained.

So here, perhaps, is a Holmesian series of inferences with the central ones knocked out: *the existence or imminent existence of an Anthropocene age reduces the likelihood that effective action will be taken to mitigate the bad effects of climate change.* I doubt very much that Holmes or Watson or their literary agent would be much concerned with climate change, given the age in which they lived. Holmes's embracing of nicotine and cocaine, his patronizing attitude to women, his nauseating racial stereotyping, indeed most of his attitudes and concerns, were standard for the Victorian society that he loved and fought to preserve.* On the other hand, the idea of an Anthropocene age would have struck a chord with the celebrated detective, living as he did in the midst of a London that was the workshop of the world, in an England at once blighted and boosted to prominence by the first and second industrial revolutions.

It might initially seem to be a little counterintuitive to claim that, because humankind can and does greatly influence the physical world on which we live, we are less likely to take action that reduces one of our worst influences. This contrast between starting point and final inference is just the sort of thing that Holmes thrived on, of course. Here are the missing links that connect the two ends of the inference chain. First, the very existence of an Anthropocene age means that, by definition, our species has influenced the physical world enough to change most of it visibly. Second, it is reasonable to assume (it is plainly the case) that these changes have been driven by economic activity. That is, preindustrial societies have not contributed much to the Anthropocene, and industrial societies have become so due to economic forces—people want to be better fed, to be better housed, and to be richer. Third, the economic development of the world is unpredictable (see chapter 34). This unpredictability is either inherent or simply beyond our ken at the present time—it doesn't matter which.

Thus the first three links lead us to conclude that human economic activity has changed the face of the Earth unpredictably. Put another

---

*In fact, Holmes and Watson were rather inconsistent in their attitudes toward race, sometimes espousing acceptance and tolerance ("The Yellow Face") and sometimes reduced to cringeworthy racial slurs ("The Three Gables").

way, climate models that seek to predict the state of the Earth in, say, the 2060s cannot do so without accurately modeling the development of world economics between now and then. Will China continue to power its industrial development with dirty coal? How fast will India develop waste management infrastructure? How will the rate at which Indonesian rainforests burn be influenced by the economic development of that country (and vice versa). How will the availability of Russian gas be affected by geopolitics, and how will it change the balance of fossil fuel consumption around the globe? You see how complicated and nonlinear the whole issue becomes. To predict world economic development decades ahead, we need to predict world political developments decades ahead. There is no algorithm on God's green Earth that will ever be able to do that. (If you still doubt that economics is unpredictable, read on. Chapters 34 and 35 should convince you.)

So, finally, the fourth link. Because the way we influence our planet's geography and ecology is both profound and unpredictable, we cannot believe detailed climate model predictions for future decades. Yes, trends can be foreseen because of the long timescales involved with physical climate processes such as ocean heating, but these trends may be swamped (this is a prediction, ironically) by short-term stuff we didn't see coming.

Fifth—the weakest link in the inference chain, I admit—such unpredictability gives wiggle room to climate naysayers and doubters, and weakens resolve to make costly changes for the better.* Hence, the fact that we are greatly influencing the Earth means we cannot predict how things will change, because we cannot predict ourselves. We would have a better shot at predicting the natural evolution of the climate if we didn't affect it so much.

Elementary, my dear Watson.

---

*Here is a historic example of how different levels of doubt about the future modify resolve to prepare for it. In the late 1930s both the United Kingdom and the United States were developing radar systems to detect approaching enemy aircraft. The British were damned sure that war was coming, and soon, and by 1940 had developed a functioning radar system that was integrated into their air defense command and control structure. The Americans pushed their radar development less urgently because they were less certain they would soon be at war. Pearl Harbor was a nasty surprise made worse by an ineffective radar system, in contrast to the Battle of Britain, which was won largely by radar. See Denny (2007a) chapter 1.

(34) Ferguson versus Krugman

Ding ding, seconds away, round 137, a bare-knuckles slugfest between, in the blue corner, "Punchy" Paul Krugman and, in the red corner, "Knockout" Niall Ferguson. It would be funny if it weren't so serious.

This is a heavyweight contest, no doubt about that. Niall Ferguson is the Laurence A. Tisch Professor of History at Harvard University, a Senior Fellow of the Hoover Institute at Stanford University, a Senior Research Fellow at Jesus College, Oxford University, author of many books, and articulate presenter of half a dozen acclaimed TV series. The pugnacious Scot, an economic historian, spreads his conservative views and lambasts his opponents and detractors online and in print via the pages of the *Daily Beast*, the *Huffington Post*, and *Newsweek*. Paul Krugman is Professor of Economics and International Affairs at Princeton University, Centenary Professor at the London School of Economics, a Nobel prizewinner in economics, and author of over 20 books. The New York native espouses his liberal views on all things economic and political in the pages of the *New York Times*.

If you and I were to get into a seedy barroom brawl with these two fighters, we would end up dead: Ferguson would smash my skull with a club, while Krugman would slip a stiletto under your ribs. The most unlikely aspect of this violent fantasy is not the undignified seedy barroom setting, nor the mindless bloodletting, but rather the notion that the two of them could ever possibly fight on the same side.* If we were to get into a debate about economics with either of these giants, then needless to say, we would lose ignominiously. Each of them would counter any feeble observations we could put forward, and provide telling evidence in favor of his views on the matter under discussion. We would leave the debating chamber both humiliated and convinced of the rightness of the arguments presented, and of the wicked wrongheadedness of the other giant. To be specific: Ferguson believes that austerity is the only way for nations to emerge successfully from the

---

*Ferguson and Krugman have spilled a lot of ink disagreeing with one another and decrying in no uncertain terms the underhanded and unprofessional antics of the other. See, e.g., Benko (2013), Cassidy (2012), Ferguson (2013), or Ferro (2015).

recent financial meltdown, whereas Krugman believes that austerity will do harm and that nations should spend their way out of the troubles caused by the meltdown. Save or spend—is there anything more fundamental in economics?

Say I am driving a nation's economic car, and these two are my navigators. We reach an economic T-junction; Ferguson says turn right, whereas Krugman says turn left. What should I do? If Ferguson were my only passenger, he would convince me about the *right* path to take; were Krugman my sole navigator, I would be *left* in no doubt about my route. But here I am with both of them on board, stuck in the middle of the economic road going nowhere. Of course, the point of this little *gedankenexperiment* in navigating the complex economic highways and byways is that these guys are world-renowned experts, and yet they cannot agree about something as fundamental as left or right / spend or save. Surely if highly educated, intelligent, and experienced experts disagree so strongly about the elementary basics of economics, then the only conclusion we mere mortals can come to is that *nobody* understands the subject. Or perhaps one of them is right, but nobody else understands enough to recognize that he is right. Whichever is the case, we can safely conclude that, as a species, humans do not understand economics. Therefore, we are unable to predict its course. Let us be generous to our expert economists (not just Ferguson and Krugman) and say that some of them are capable of seeing a short distance ahead—after all, the fiscal farce of 2008 was foreseen, in part at least, by a few insiders—but *nobody* can see ahead decades, so as to predict the world economic scene in the 2060s, say.

35 Nobody Understands Economics

Perhaps the last couple of sections haven't completely sold you on the notion that (a) nobody understands economics. It is an idea that is central to what I have to say about the way things are going to develop. Add to it another idea—hardly a new one—that (b) different groups of people can't cooperate for the common good. The latter notion is developed in the next few sections; it combines with (a) to sentence us all to a degraded future—degraded compared to what is possible. The perpetual economic mess of the world brings to mind an image of noisy babies in a playpen with their stuff strewn around haphazardly: food, toys, feces. In this section I will reiterate the view that economics is beyond our ken, in the sense that we have no idea how it will develop long term, and will hammer home the central point that, as a consequence of (a) and (b), the world is getting worse. That is, in a number of meaningful ways, the world in which your grandkids live is going to be a noticeably less pleasant one than ours, and the world that their grandkids inhabit will be noticeably worse again.

Not fun stuff. From the writer's point of view, this is the worst kind of prediction to have to make. Suppose I was a Doom 'n' Gloomer—then I could at least entertain you with some striking prose about a fiery hell of our own creation consigning billions of innocent children to short brutal lives and early painful deaths, or some such—it would be dramatic fare, an apocalyptic page-turner. At the other extreme, as an adherent of the optimistic branch of Technofix, or a self-satisfied LPG groupie, or any kind of BaU advocate, I could prophesy a beautiful future for humankind that would leave you with a warm fuzzy feeling inside; healthy happy kids with iPads and healthy happy puppies skipping through fields of daffodils—a feel-good page-turner. But instead, I find myself having to convince you of the least interesting possibility, one that is at once unhappy and dull. (Hence, perhaps for palatability, I've given this book some schtick.) Sorry about that, but at least this viewpoint has the virtue of being the most realistic.

＊　＊　＊

It has been said that "an economist is an expert who will know tomorrow why the things he predicted yesterday didn't happen today."* Perhaps you disagree. Perhaps a Dubious, Unconvinced, Doubtful, or Exasperated reader (henceforth, a *dude*) might pipe up and ask how I can claim that the world is perpetually in an economic mess.

Dude: How can you claim that the world is perpetually in an economic mess? We have had good times as well as bad; economies are cyclic.

---

*Laurence Peter, *The Peter Principle* (1969). We will hear more from Dr. Peter later.

Author: Well, for as long as I can remember, probably for much longer, and certainly for as long as we have all lived in a globalized economy, there have been billions of people who are hungry and without the basic needs of life. This observation is not meant as a shallow political remonstration about how we should all care for one another, but rather, it reflects on the fact that economic practices don't work always for everyone—and don't work at all for many people. Even within developed countries there is a significant underclass of people who are economically excluded through no fault of their own. The political choices of a nation may alter the size of this underclass, but it is the nature of capitalist economics for there to be an underclass.

Dude: Much of the mess we find ourselves in is due to the fact that people have enemies—it's human nature, not economic misunderstanding. Even as you write these words, there is a sad tragedy unfolding in Paris, with some radicalized fanatics killing innocents. You're putting too much of the blame for the mess of the world on economics.

Author: Not really. Our inability to get along is certainly an important factor in shaping our future, as I will argue later, but much of the reason why nations don't get along is economic. Anyway, it is the combination that counts. It is a mistake to think we understand economics when we don't, and to make predictions based on perceived economic progress.

Dude: What makes you so sure that nobody understands economics? I get that two experts disagree on fundamentals, and that this tells us nonexperts that the subject is not fully understood. But two meteorologists might disagree over whether it's going to rain tomorrow, and still have a good understanding of their subject.

Author: Economics is not like meteorology. In meteorology the uncertainties are due to chaotic dynamics or random effects. These exist in economics too, but economics is not underpinned by a basic understanding of the laws. The laws of economics are not physical laws that can be described accurately by mathematics. What underpins economic activity—the basic unit of economics, if you like—is a person who buys or sells based on panic or fashion as well as on rational self-interest. Yes, *some* economic principles are understood:

lowering interest rates stimulates investment, for example. *Some short-term linear predictions can be made:* the Chinese economy has been growing for three decades and will continue to grow, at least for a while. But these linear predictions don't require deep knowledge and are short term, as we saw in chapter **30**. Nobody has a handle on economic prediction into the hazy distance, such as what the world's economy will look like in the 2060s.

Dude: At the risk of you turning me into a straw man, let me ask this. Can you give me something more to convince me that nobody understands—more than the profound disagreements between Ferguson and Krugman?

Author: I am glad you asked that question, Dude. Krugman—a Nobel laureate in economics, let me remind you—recently wrote an article entitled "Nobody Understands Debt." (It has, of course, been criticized in the business journals.) Ben Bernanke of the US Federal Reserve has been quoted as saying that nobody understands gold prices, including him. A prominent American financial journalist, Michael Lewis, said in a recent interview that nobody understands the stock market, and that its complexity is a source of instability. A prominent British economics journalist, Jeremy Warner, wrote a commentary entitled "No One Really Understands What's Going On in Our Economy." He notes that contradictory data (such as a million new jobs, yet no signs of real economic growth) are "about as useful as a chocolate teapot." What meteorologist would say that about weather data?*

Dude: Climate models can make accurate predictions based on the physics—you said so in chapter **15**. They can account for the uncertainties due to economics by postulating different scenarios—say different rates of GHG emission up to year 2070, due to different economic trajectories for the world between now and then. These scenarios are a bit like the *ensemble forecasting* of meteorologists: they know that they can never measure current conditions accurately

---

*Articles of this type, in which experts confess their own ignorance and that of every other expert in the field of economics, are more common than you might suppose. See Heaven (2013), Keen (2015), Krugman (2015), Warner (2013), and the Youtube video *Michael Lewis: Nobody Understands the Stock Market*. See also Hazlitt (1988), who shows that expert cluelessness is not new.

enough to make predictions, because the underlying dynamics are chaotic and thus depend very sensitively on initial conditions. So they start with a spread of initial conditions and see what they lead to—a spread of predictions about what the weather is going to be. That's why we get statistical weather reports ("70% probability of precipitation tomorrow"). Aren't climate models accounting for the uncertainty in economics in a similar way, with different scenarios?

Author: A fair point, Dude, but the range of scenarios is small, and the space of possibilities they are attempting to cover is very large. Meteorological models are based on a detailed statistical understanding of what the spread in any given measured parameter will be, and therefore can account for it (70% probability of precipitation, in your example). Yes, climate models can do the same thing for the physical part of their predictions, but not for the economic part. They don't give probabilities that a chosen economic scenario will arise.

Dude: Ah. So you're saying that a climate model prediction for 2060 is right—within the physical uncertainties—if the assumed scenario is right, but that we have no way of quantifying how likely that scenario is, because we know jack about what the economic/political/ human situation will be like then?

Author: Exactly. And this uncertainty about climate models leaves wiggle room for climate inaction which, if the climate models are right (see the doubt?), may be disastrous.

Dude: Wow, you're brilliant. I'm impressed! I must recommend your book to all my friends.

● ● ●

So why do (a) and (b) sentence us to a degraded future? The Sherlock Holmes argument of chapter 33 suggests a lessening of resolve among people because of doubts about climate change—due to the uncertainty of predictions, among other things. These doubts were reflected in a pre–Paris Summit opinion poll (late 2015) in which two out of three Americans said they accepted the reality of climate change, with the vast majority accepting that humans were at least a part of the cause of it. Yet only one out of three was even moderately worried about climate change; the rest were not especially concerned, because of more immediate issues to worry about, such as terrorism or jobs.* Combine this diluted resolve to do anything, or at least to do anything costly, with the uncomfortable fact that not every nation will be affected equally by climate change—in the short term there will be winners as well as losers, as we see in the next section (chapter **36**)—and we can anticipate that humankind is being pulled in different directions, according to varying circumstances. Self-interest usually wins out over the collective good, a phenomenon formalized in the mathematics of the game called the *Prisoner's Dilemma*, where two prisoners have to decide if their interests will be better served by betraying each other or remaining silent. More narrowly, a version called the *Tragedy of the Commons* applies to natural resource exploitation. Both of these phenomena are explained nonmathematically in chapter **37**. The insidiousness of these notions is that each group or nation may be acting rationally when being selfish; they may be aware of the Prisoner's Dilemma phenomenon, and be aware that the long-term result of improved cooperation will be better for everyone, yet still act selfishly because they do not trust the other groups or nations.

So we can expect a slow, muddled, and incomplete response from humanity as a whole in response to the pressing challenges of climate change. History backs up this somewhat depressing notion in spades, as we will see. I will look at one example in more detail, because it pertains to climate change; deforestation is the subject of chapter **38**. Thus (a) and (b) alone give cause for concern about our collective ability to head off the worst effects of climate change. Add to these another human characteristic, (c)—collective stupidity, examined later on—and we end up with my pessimistic view of the road ahead. We have the smarts to think up technofixes to mitigate or even reverse the bad effects of climate change, but collectively we are too stupid to put them into effect fully, or in a timely manner.

 **Winners and Losers**

The really insidious aspect of climate change is that, initially, there will be winners as well as losers. I do not mean simply the oft-repeated

*Borenstein (2015).

broad statement that we have all borrowed from our descendents, leaving them our mess to clean up, so that we are the winners and they are the losers of climate change. Such a case—appalling enough—would at least be geographically equitable. The uncomfortable fact is that the effects of climate change are distributed unevenly across the world and that some regions will come out ahead, at least in the short-to-medium term. That is, some parts of the world (some nations) will benefit from the initial changes in climate—from global warming—that our industrial revolutions have initiated. This aspect is insidious because it is unfair and provocative and, more importantly, because it will dilute our global will to address the problem. Local selfishness will win out over the common good—a human characteristic that I examine in chapter 37; here I outline the unevenness of short-term climate changes around the world.

Short-term effects are predictable, as you may recall from chapter 30—it is the long-term predictions of climate models (several decades hence) that are doubtful, due to economic uncertainties. So, here are some short-term predictions of what will very likely happen as a result of the world warming a couple of degrees. The natural world will change—there will be winners and losers. Polar bears, Bengal tigers, Arctic foxes, woodland caribou, white-fronted lemurs, cuckoos, and grouse—these animals are among those species that will lose out. Merriam's kangaroo rat, greater yellowlegs, American bullfrogs, some Antarctic penguins, and many insects (including forest pests normally killed in winter) will benefit. The chinstrap penguin will initially benefit but then lose out. This list is very incomplete, of course. Losers will decline in numbers and will perhaps go extinct, or will be reduced in range. Winners will expand their range and numbers; in many cases this expansion will be a continuation of current trends. Thus in Britain, many of the animal species are at the northern end of their global range, so that increasing average temperatures will place them closer to their comfort zone. They will largely benefit from the first phase of global warming.

The same is true for humans in Britain, despite the fact that the industrial revolutions began in this part of the world. Anthropogenic carbon dioxide does not hang out near the factories that generated it—it spreads across the world. Its influence is determined by the laws of physics, not of natural justice. Most of the developed nations—those in western Europe and North America, not just Britain—are in temperate latitudes of the Northern Hemisphere, and these latitudes will initially benefit from global warming, on the whole. (Western Europe may suffer if the Gulf Stream changes course as a result of North Atlantic warming, however.)

Australia is one of the few developed regions that will suffer from the get-go as a result of global warming. Developing and undeveloped regions are often in subtropical latitudes or are equatorial (Brazil, India), and human life in these areas will deteriorate first. Low-latitude regions will warm up more than temperate latitudes and will receive less rainfall; the rain they will get is going to arrive less often and in more powerful downpours, enhancing the likelihood of both drought and flooding, with increased temperatures in regions that are already hot. Some of these effects will spill northward: the southern United States will experience more drought and more tropical hurricanes; there will be drier and more lasting droughts in Spain, Italy, and Greece (also in Australia and South Africa). Changing rainfall patterns will imperil food production in some present-day breadbaskets and create new breadbaskets elsewhere (wheat yields in northern Europe will increase some 30%).

Generally, developed countries will benefit, because of their latitude. Increased temperatures will mean longer growing seasons and greater food production. Higher minimum nighttime temperatures will permit a wider variety of crops to be grown. As crops fail in the heat of the subtropical latitudes, some of them will become viable farther north. Melting Arctic ice will free up the long-sought-for "Northwest Passage," permitting a major new shipping route between Asia and Europe.* Melting permafrost will increase the area of land that is farmable, and ease the extraction of underground natural resources. The big winners will be Scandinavia and, especially, Canada and Russia. Alaska and Greenland (where the growing season has extended by two weeks since the 1970s) will also become more

---

*Denver entrepreneur Pat Broe bought the port of Churchill, Manitoba (for $7), in 1997 and has developed rail links to it. He is betting that this port, the only subarctic industrial port on the continent, will become a major player when the Northwest Passage opens up. Its location at the western end of Hudson Bay will facilitate trade between the US Midwest and Russia and northern Europe.

temperate and richer, as the land becomes more productive and more valuable. People will want to move there, driving up real estate prices. Germany will become warmer and may have a Mediterranean climate by the end of the century. It and other nations of the temperate north will benefit from increased tourism as well as more widespread and productive agriculture. Land prices will rise in these regions, too.

Global climate predictions are more reliable than local ones, so regional effects are less easy to be sure about. Globally, sea levels will rise (estimates vary from 7 inches to 23 inches by the end of the century, depending on scenario), and this will likely cause populations to move inland, away from the coasts. This phenomenon will be disastrous for some low-lying countries such as Bangladesh, Seychelles, and many Pacific islands; it will cause less momentous population shifts in other parts of the world. Thus, Florida will lose land and will suffer from a rising water table. Within those nations that initially benefit from global warming, there will be changes in population and real estate prices. Northern US cities that currently have hard winters, such as Buffalo, will increase in size as their winters mellow, as people in Houston abandon their increasingly humid city, and as citizens of Arizona and Nevada leave their desiccated states. The general rule is that low-latitude property prices will tank, whereas temperate latitudes will become the new Sun Belt. Developed countries will be able to adapt to sea-level rises by constructing costly levees and drainage infrastructure, whereas undeveloped or developing countries will suffer more because they can adapt less, thus increasing inequality.*

Immigration to Canada will increase as the range of habitable land moves northward—in fact, Canadian population growth will be among the highest in the developed world. The geographical center of population in North America will drift northward. That is, the American South and Southwest will lose people as it warms and become drier, whereas the northern parts of Canadian provinces, and even the northern territories, will gain residents. Today there are 9 Americans for

---

*An example of the influence of wealth on adaptability is found in mortality figures that are due to increased temperatures. For a given rise in temperature, many more Indians die than Americans, because the United States has more widespread resources such as air conditioning and refrigeration, to mitigate the effects.

every Canadian, but this numerical disparity will reduce markedly over the next 50 years.* Vast tracts of Siberia, under snow for millennia, will reveal soil undepleted by agriculture. Russia has the potential to become an economic superpower as it warms in the south and thaws out in the north.†

Freshwater shortages may lead some countries to fail or may lead them to war. Bad luck in geographic location is causing severe water shortages in North Africa and the Middle East (the latter has 5% of the world's population but only 1% of its fresh water), for example in Egypt, Libya, and Yemen. China has extensive problems with insufficient supplies of water for drinking and agriculture, as have Australia and the American Southwest.

The opening of a Northwest Passage and the warming of the Arctic (hence making underground resources more accessible) may lead to a land grab in the region—already there is an uptick in activity there and feisty jockeying for position among the five contenders—Canada, Denmark (which administers Greenland), Norway, Russia, and the United States. There may be similar land grabs in the Antarctic. Economic migration on a large scale will occur from stressed or failing nations in the tropics to higher latitudes; from Central America northward, or from Africa to Europe. These migrations will only make worse those that are happening in these regions today for other reasons. That is, distressed peoples moving north to Europe or the United States to escape war or poverty today will be joined in future decades by those obliged to do so by the effects of climate change—failed farms, failed infrastructure, failed states. Perhaps Brazilians will look covet-

---

*The disparity has decreased slightly over the last 50 years; in 1965 there were 10 Americans for every Canadian.

†Note the cautious wording here—it takes more than a favorable climate to generate a powerhouse economy, and Russia has no history of surehandedness in this department, to say the least. It has been claimed that Russia will gain more from the first phases of global warming than all other nations and that, as a consequence, Russian leaders *want* climate change and are happy to drag their feet at climate change meetings that call for binding commitments. If true, we have here a significant example of selfish behavior, of the type I will discuss in the next section. If untrue, we see in the allegation a glimpse of the discord and suspicion that the variable effects of climate change induce. See Easterbrook (2007).

ously at the cooler regions of northern Argentina, or Pakistanis in the sweltering lowlands will envy the milder temperatures of Afghanistan—a region with which they have strong historical ties. Inequity will further increase beyond that caused by globalization, due to the unfair effects of variable global warming, and this inequity can be expected to stoke regional disputes. There will be squabbles over GHG offsetting and other mitigating effects, which will only be exacerbated by the winner/loser aspect of it all. Such geopolitical adjustments cannot be scientifically predicted in detail because geopolitics is not science, but a glance at history shows that they are in the cards.

Lesser climate changes than the one we are about to undergo have had significant effects on human history. Thus the Mayan empire in Central America probably declined due to drought, which reduced agriculture and available drinking water. On the other hand, the Medieval Warm Period (the 300 years from about 950 AD, during which the North Atlantic region was relatively warm—though, note well, not as warm as today) benefited some European countries. Spain, France, and England increased their populations and the size and number of their cities, as well as their agricultural output, during this time. This is not a coincidence. Trade routes expanded, along with the quantity and variety of commodities and products that were bought and sold. Given that global warming this century will embrace both these types of change—droughts and increased mid-latitude temperatures—we can anticipate a consequential decline or growth among many of the nations of the world.

One observer has noted that it is a "dangerous avenue to say there are benefits from climate change," perhaps because it will hinder international attempts to find a solution, or will reduce its effects.* It is also deceptive, because in the long run, we will all be losers. The changes that are going to benefit mid-latitude regions will later on become too much of a good thing, if temperatures rise much more than the 2°C guardrail discussed in chapter 17. For a rise of 3°C–4°C, climate physics tells us there will be much more marked changes to precipitation patterns than arise at lower temperatures; these will

---

*Anders Portin, of the Finnish Forestry Industry Federation, quoted in Doyle (2007).

eventually outweigh any benefits that mid-latitudes initially reap from warming.*

•   •   •

The Prisoner's Dilemma is a well-established hypothetical phenomenon to which we now turn. Applied to the current world, poised as it is on the brink of the Anthropocene, with uneven consequences anticipated for the coming climate change, we can expect little in the way of concerted action to avert the worst effects, even though every nation may know it is in their best interests to do so. This theme—the human inability to act collectively for the common good—is explored in the next few sections. It underpins my somewhat negative view of the future but does not define it; it takes the shine off the better world that might be embraced if we could learn to love our technological monsters, a view I will argue for in the penultimate chapter of this book.

37 The Prisoner's Dilemma

Often cited and equally often badly explained, the Prisoner's Dilemma is a mathematical exposition of selfish behavior. Understood properly, however, it is more than that; it explains why even rational people may choose to act selfishly, even though they know the consequences will be worse than if they cooperate with others and act for the common good. The Prisoner's Dilemma is the poster child of game theory, a branch of mathematics invented by the mathematical physicist John von Neumann in the 1940s, and applied almost immediately to economics by him and Oskar Morgenstern. Its significance in economics can be gauged from the fact that, of all the winners of Nobel prizes for economics, 11 have been expert in game theory.†

*For more details on the winners and losers of climate change, see, for example, Adger and Paavola (2006), Doyle (2007), Easterbrook (2007), Evars et al. (2010), Lean (2015), Palmer (2012), and Sim (2015).
†Including John Nash, who won the 1994 prize along with two other economists, who were also game theorists. Nash was the subject of the 2001 movie *A Beautiful Mind*.

Here is the classic case. Two nefarious individuals—let's call them Knuckles and Spike, members of the 1940s B-movie criminal milieu—are arrested for a crime and held in isolation. Each is told that if he rats out his colleague, he will go free and his colleague will get three years in the pokey. If they rat out each other, both will get two years. If neither sings, they will each receive a year in jail for a lesser charge. What should they do?

It is in the best interests of both if they remain silent—if Knuckles says nothing and trusts Spike to do the same, and vice versa—for then the pair serve a total of only TWO years behind bars (breaking rocks, let's say). If Knuckles keeps quiet but Spike betrays him, or the other way around, then the pair get THREE years penal servitude between them. If they both sing, the pair will be employed by the government preparing roadbase for a total of FOUR years. But—and here is the rub—the interests of the individuals are not the same as the interests of the pair. Knuckles's best strategy is to rat out Spike if he doesn't trust him to keep his trap shut. For Spike, likewise. It turns out that, perhaps surprisingly, both of these individuals are au fait with game theory and so they know that cooperation and trust—both remaining quiet—is the best strategy. They both know that the best result for Knuckles as an individual (and for Spike) is to talk, so Spike (Knuckles) also talks, to lessen his sentence. It is a question of trust: the group (here the pair of criminals) is better off if its members trust each other, but being human they don't, so each member acts selfishly, to the detriment of all.

There are unstated assumptions in my explanation: the game (the offers made to Knuckles and Spike, and their responses) happens only once, and there is no follow-up, such as reprisals. When the game is played many times in succession, the outcome can be different; for example, the two prisoners can learn to trust one another. The game can be extended to more than two players. If there are three players in the Prisoner's Dilemma game, which is repeated, then perhaps Knuckles and Spike learn to trust each other but neither trusts Filthy McNasty. Optimal strategies build up as the game is repeated. The same basic idea is ever present, however: lack of trust leads to rational behavior that is selfish and not optimal for the group.

The real world is more complex than this simplified mathematical game, you will be unsurprised to learn. Thus, the two-person game between Knuckles and Spike was entirely symmetrical between the players. However, many real-world applications are not. For example, in the real-world Climate Change Negotiations game, some players have developed their economies by burning fossil fuels, whereas others are trying to conserve fossil fuels; some players are bigger than others; some players have more at stake than others.

In the Advertising game, two companies with similar products compete for market share and profits. If neither advertises, they split the market 50/50. If one advertises, then that company increases its share and profit, despite advertising costs. Both end up advertising, and split the market share, though profits are lower because of advertising costs—the companies both lose profits, because each cannot trust the other not to advertise. There are many other economic applications of the Prisoner's Dilemma. The day I write these words, an article in the Canadian press covers the travails of the considerable mining sector in that country. Because of the slowdown in China, mining sector profits are squeezed. It would make sense for each iron ore mining company (to take one example) to reduce production, thus restricting supply so that prices will naturally rise. However, the companies don't trust each other, so each feels obliged to continue with maximum production so as to avoid losing market share, thus keeping prices down and their industry depressed. Much of the sluggishness of the world's response to the financial crisis of 2008 has been analyzed in terms of game theory in general and the Prisoner's Dilemma in particular. Short-term selfish actions dominate; effective mechanisms for cooperation and coordination do not exist, and if they did, they might fall foul of regulations aimed at curbing price fixing. Applying game theory to ongoing climate change talks leads some commentators to be pessimistic about the outcome.

When even the math is against you, it is hard to be trusting in this cold, cruel world.*

---

*The Prisoner's Dilemma is explained clearly in a YouTube video of that name; it is invoked or applied to real-world situations in two telling examples by McGugan and Younglai (2015), and Weldon (2015a). See also the *Economist* (2007) and Rehmeyer (2012) for game theory applications to climate change.

* * *

The Tragedy of the Commons is a Prisoner's Dilemma variant that is all too often applicable to climate and environmental matters. In this case, rational individuals acting independently and selfishly behave contrary to the best interests of the group to which they belong, by depleting a common resource. These last words, "by depleting a common resource," characterize the tragedy of the commons. The phrase originated in England in 1833, when an economist called William Foster Lloyd observed the effects of unregulated livestock grazing on common (unenclosed) land. Farmers would put their sheep or cattle out to graze on common land ahead of their own private land, so that the public resource (grass) was depleted before their own private reserve. All farmers felt obliged to do this, or else they would miss out on free feed for their animals. The result was overgrazed common land, which was then of no use to anyone. In 1968 the notion was updated by the American ecologist Garrett Hardin, and since then the tragedy of the commons has been observed in many ecological instances.

For example, the Grand Banks cod fishing grounds have been visited by fishermen of many nations for centuries. Technological advances meant that fish catches were greatly increased during the 1960s and 1970s, resulting in an international feeding frenzy as more and more cod were taken; by the 1990s the Grand Banks were depleted of cod, to the extent that they were no longer economical to fish and the cod population was so low as to be endangered. A common resource had been depleted; more-restrained fishing (perhaps via regulation) might have kept the cod stocks high enough indefinitely. In the fate of Martha the pigeon (chapter 4), we have already seen a similar case whereby individuals sacrifice a common stock and the well-being of their group for selfish gain. Had passenger pigeons been culled at the rate which they bred—an annual bounty of free meat—they would still be with us and we would still be benefiting from them, and from the foresight of our forebears. But our ancestors chose to look after themselves and not us, so passenger pigeon pie is not on our tables. Here is a clear example of borrowing from future generations—very clear to us because, in this case, we are one of the future generations that lost out.

Note that in both these cases—Atlantic cod and passenger pigeons—a near infinite future resource (all the future generations of cod and of passenger pigeons) has been squandered for a finite immediate gain. Today there are many examples of the tragedy of the commons in which we are the selfish and short-sighted squanderers, robbing future generations of, for example, California groundwater, clean air, unpolluted oceans, and trees, among other things. The case of deforestation and unregulated logging is interesting, because it has not yet led to a total depletion of the world's forests, and might not ever do so, due to a certain level of foresight and action, plus the particular dynamics of deforestation and reforestation. Because it is a little different, I will look into deforestation (as a tragedy of the commons) in more detail in the next section, chapter **38**.

One particularly scary example—as yet only a potential tragedy of the commons, but one that may be realized within the next few decades—concerns antibiotics. Since the introduction of penicillin in 1928, antibiotics have saved millions of lives and prevented disease in millions more people and in livestock, thus playing a double role in increasing the human population and keeping it healthy and fed. Inexpensive antibiotics became widely used; consequently, bacteria around the world developed resistance to different antibiotics, rendering them less effective. New antibiotics have been developed over the years, resulting in a kind of arms race between pharmaceutical researchers and bacteria; as bugs became resistant to older antibiotics, so new antibiotics are developed. The phrase commonly used is an "antibiotics pipeline"; new drugs emerging from the pipeline keep bacteria in check even as the older drugs lose their efficacy. The trouble is, the pipeline has been drying up for some time. Public health officials in the United States, the United Kingdom, and the World Health Organization have been sounding alarms. "We're on the precipice of returning to the dark days before antibiotics enabled safer surgery, chemotherapy and the care of premature infants," according to Dr. Helen Boucher, an infectious diseases expert at Tufts Medical Center in Boston. "The world is heading towards a post-antibiotics era in which common infections will once again kill. If current trends continue, sophisticated interventions like organ transplantation, joint replacement, cancer chemotherapy and the care of preterm infants will become more difficult or even too dangerous to undertake," says Margaret Chan of the WHO.*

The coming antibiotics apocalypse, as it is increasingly being called by the press, fits the definition of a tragedy of the commons. Here the common resource is the ever-growing suite of antibiotics. Unfortunately, the rate at which new antibiotics come down the pipeline is slowing to a crawl and is slower than the rate at which bacteria are developing resistance. Selfish action by individual doctors, farmers, and nations in overusing antibiotics has led to the increased resistance. Antibiotics have been widely distributed to patients by their doctors simply because the patients requested some medication, rather than because it was strictly necessary on medical grounds. This malpractice has been widespread in the United States and other developed countries, but has been reduced or eliminated over the past few decades—for example, the problem began to be addressed in the United Kingdom as long ago as 1970. Currently, the overuse of antibiotics in animal husbandry has been blamed for the result of recent research in China, which demonstrated that our most powerful antibiotic (colistin) is losing its effectiveness against a type of bacterium found in pigs.

A potential tragedy of the commons can be addressed only by common, prompt, and concerted action by the group of people who are affected—action directed against the selfish exploitation of the common resource, or action directed to avoid its depletion. In many cases such action has been absent (we cannot resurrect huge flocks of passenger pigeons), or has been too little, too late (Atlantic cod still exist, but it will be several generations before they can be fished again). It is to be profoundly hoped that concerted action may yet avoid an antibiotics apocalypse.†

●   ●   ●

*Boucher is quoted in Steenhuysen (2013); Chan is quoted in Ridgewell (2015). For antibiotic-resistant bugs, see Sun and Dennis (2016).

†I draw this account of the tragedy of the commons from the following articles: Gallagher (2015), Hanlon (2013), Hardin (1968), Ridgewell (2015), and Steenhuysen (2013).

Garrett Hardin started off his 1968 paper about the tragedy of the commons by saying, "The . . . problem has no technical solution; it requires a fundamental extension in morality." How likely is that, do you suppose?

 Deforestation

There are about three trillion trees in the world. Roughly half of these are in the tropics or subtropics, a quarter are in temperate regions, and the remaining quarter are boreal (northern). About 25 countries (plus a number of small island nations) are more than half covered in forests. Forests cover some 26% of the land area of Earth—the numbers are uncertain and vary with source, depending on how we define a forest. Two-thirds of forest area is managed. If we chop up the forests (figuratively, I hasten to add) by continent, about 12% of North America is forested, along with 17% of Africa, 19% of eastern Asia, 25% of South America, and 27% of Europe and western Asia. Russia, Brazil, Canada, and the United States among them contain about 19% of the world's forests. Forest coverage, and the annual loss in forest coverage, can be estimated and monitored globally by satellite.

Phytoplankton—microorganisms that live in the sunlit upper levels of the world's oceans—account for half of all photosynthesis, and forests account for most of the rest. Tropical forests contain a quarter of all the carbon that is found in living things, and it is the carbon content of forests that gets the world's attention. Environmentalists and conservationists are concerned about deforestation because they like trees; climatologists are concerned because of the important role trees play in determining the GHG content of the atmosphere. A significant part of the *carbon cycle*—the movement of carbon around the biosphere and the Earth's crust—is the loop between photosynthesizing organisms and the atmosphere. The process of capturing carbon from the atmosphere, by absorbing carbon dioxide, is called *biosequestration,* and forest trees do a lot of it, via transpiration. About 30% of anthropogenic carbon dioxide emissions are absorbed through the leaves of trees (another 30% is dissolved in the oceans). Consequently, cutting down trees on a large scale leads to an increase in atmospheric carbon dioxide; the Intergovernmental Panel on Climate Change estimates that 20% of GHGs are due to such deforestation—this is more than the total tonnage of CO_2 added to the atmosphere each year by emissions of all road vehicles (about 15%).

The number of trees varies over time due to many factors, natural as well as manmade. Thus storms, infestations, and heat waves have all killed trees in large numbers during the brief periods of time that these natural phenomena occur. Carbon sequestration by European forests was reduced by a third during the severe winter storms of 1999 as trees were knocked down like ninepins (and timber prices consequently dropped by a half while these windfall supplies lasted). An ongoing pandemic of bark beetle infestation in British Columbia, Canada, has destroyed some 435 million cubic meters of timber since 2004. Forest fires that resulted from the record 2010 heatwave across Russia destroyed 23,000 square kilometers of forest. All these natural disasters are local in space and time and so of limited global significance. The most significant natural cause of deforestation is due to El Niño / La Niña, because of its global reach. More important than any natural causes of deforestation, however, are the actions of humanity.

About 50,000 square miles of tropical forest were cut down during the decade from 2000 to 2010. The biggest single reason was conversion of forest to cropland and pasture—mostly slash-and-burn subsistence farming, but also industrial-scale cattle ranching and (in Brazil) soybean production. Logging—legal and illegal—also takes a significant toll. Some of the losses of forest due to human activity are unintentional, due to fires and overgrazing, but most are intentional, driven ultimately by poverty and profit. Apart from contributing significantly to atmospheric GHG via loss of biosequestration, clearing tropical forests also endangers species due to loss of habitat and alters water drainage and runoff.

The most deforested parts of the world are in Brazil and Indonesia. In 1998 Brazil was losing Amazonian rainforest at the rate of 7,700 square miles per year; today it is down to 2,300 square miles a year. At the turn of the millennium, 25% of GHG increase was due to forests being turned into farmland; today the figure is 12%. The rate of deforestation is slowing down worldwide, but it won't cease so long as there is money in it for a lot of people. A characteristic *forest transition curve*, a plot of tree numbers versus time, describes the change of forest cover in countries that go through a phase of deforestation during economic development. The initial tree coverage is typically high—it is the natural forest cover of a country prior to human actions—say 90%. Congo is currently at this stage, as it begins to exploit its forest resources. Then for some decades the forest cover decreases rapidly, as the trees are cleared and either burned or turned into timber; Brazil and Indonesia are on this downward slope of the curve. Then the forest coverage increases somewhat, as reforestation of regions that have lost their trees occurs, or as land that was never forested is planted with trees. These new forests tend to be monocultures that do not perform as well from the ecological point of view, in terms of providing habitat, but they do fulfill the climatological function of sequestering atmospheric carbon. India is at this stage of the forest transition curve. Finally, a new equilibrium level of tree cover is reached; Costa Rica is there. The new equilibrium level is lower than the original natural level of forest cover—perhaps 40% or 50% of the land originally covered with trees. Thus the forest transition curve of a developing country typically starts as a flat line that then dips sharply to a minimum, before increasing and flattening at a level lower than it began.

Every aspect of this subject is complex. Thus, the drivers of GHG levels due to changing forest size are not well understood. Until recently, for example, it had been assumed that the increasing concentration of carbon dioxide in the atmosphere would naturally lead to increasing sequestration as forests absorbed the extra carbon and turned it into wood, resulting in trees growing faster. In fact, more carbon is indeed captured by forests as atmospheric CO_2 levels increase, but not due to individual trees growing faster, we now know from tree-ring studies. Rather, the extra carbon is captured due to an increasing density of trees in the forests—the *number* of trees is increasing. The governmental and industrial management of forest clearing and of reforestation are complex; legislation and regulation do not always work as intended (see chapter 39). A recent research article has claimed, because of this insufficient understanding that exists at all levels, that "today's forest management is more of a gamble than a scientific debate."*

*The quote is from Bellassen and Luyssaart (2014). For more details on deforestation, how it has changed, and how it has changed the world, see also Amos (2015),

And so we come at last to the reason for this section of the book, why deforestation is included as an example of human mismanagement / inability to cooperate for the common good / tragedy of the commons. First, it is a clear example of a common good being eroded by the selfish actions of individual people (or companies or countries) for their own benefit. The long-term consequences of deforestation must be obvious to many of those whose actions contribute to it, yet in their minds the short-term gains outweigh any concerns they may have about long-term effects. Second, and more important for my purposes, deforestation has an endgame that, in my opinion, more closely resembles the endgame of the wider Anthropocene debate than do other outcomes. A Doom 'n' Gloomer would foresee the extirpation of Earth's forests; after all, if we continue burning or cutting down trees at the present rate, even if it is reduced from the rate of 15 years ago, there will be no forests left a hundred years hence. Our collective stupidity and individual greed will doom the forests, and by extension will doom us, just as it doomed the people of Rapa Nui to chop down the last of their trees—except that it didn't (see chapter **20**). At the other extreme, a Love, Peace, and Granola groupie might preach and even believe that by conserving what remains of the world's forests (how?) and planting trees, in short by sustainable forest harvesting and management, we can return to something like the pristine green Earth that existed before people arrived on the world stage and used their brains to fell trees. My analysis, soon to be summarized in chapter **41**, points to an intermediate fate for us, just like the intermediate forest transition curve; we will rebound from the destruction that we have wrought by using our foresight and cleverness, but only part way, because of our short-sightedness and collective stupidity.

S. Clark (2014), the *Economist* (2014b), Scheer and Moss (2012), and two useful websites: National Geographic's "Deforestation" (http://environment.nationalgeo graphic.com/environment/global-warming/deforestation-overview/) and NASA Earth Observatory's "Causes of Deforestation" (http://earthobservatory.nasa.gov/Features /Deforestation/deforestation_update3.php). The state of the world's forests and their importance to our future is analyzed in "Forest Health," a special issue of *Science* (2015).

The Peter Principle

Dr. Laurence J. Peter has expressed best and most precisely the "law" of social development that leads to what I call *collective stupidity* (described in chapter **40**). It doesn't hurt that throughout his life, this Canadian American educator made his point with humor—first in a best-selling book, *The Peter Principle* (1969), which has sold 8 million copies in 38 languages. The principle is that, in any organizational structure, managers tend to rise to the level of their incompetence, where they remain. Thus organizations that are large enough to require an employment hierarchy tend to be run incompetently. Perhaps the principle does not apply to small companies, which would go to the wall if they conducted business incompetently in a competitive market, but it certainly does apply to large private companies, public organizations, and government.

Public sector incompetence might be expected, because a government organization is protected from market forces and its success or failure is not necessarily measured by competitive metrics. It may well be a monopoly that is perceived to be necessary; it is often unaccountable in practice and is cut a lot of slack by its masters, who themselves are often sinecured factotums inexpert to judge success, and who are appointed according to the Peter principle. Thus, the diplomatic service of any nation can be run by-and-large incompetently because it is largely unseen and is barely accountable for much of what it does. Another very common example: gross overspending by municipal, county, state, or federal departments on public works is woefully common—I am sure you can think of an instance from your local community. Such overspending is not necessarily due to corruption; often the mistakes are simply due to incompetent individuals or committees who are spending taxpayer money, not their own, on projects that will benefit other people, not themselves or their friends.

What about the private sector? Large organizations such as multinational car manufacturers and oil companies come in for much amusing analysis by Peter. A corporation may have one capable chairperson but twenty vicechairs—many incompetent—in what Peter calls a "flying-T" formation. Corporate incompetence can be expected to be concentrated at intermediate or internal levels: surely the departments and individuals who represent the company's face in the marketplace know what they are doing? Sigh—not always. I worked for a decade designing radar systems for a multinational aerospace company that was (perhaps still is) renowned within the field as the acme of incompetence. Yet I learned that other radar corporations, outwardly more capable, also generated huge and expensive blunders. Large radar systems are very complex and require many technical people to design and build them; crucially, they are too complex for the whole system to be comprehended in detail by a single person. One internationally renowned expert in the field told me, at a systems engineering course he delivered in London in the early 1990s, that large radar systems never work as well as intended for reasons that are sociological, not technical: humans are not capable of organizing themselves well enough (David Barton, personal communication). For such hierarchical incompetence, we can thank the Peter principle.

The existence of governmental incompetence is often made very plain and obvious in a liberal nation where free speech is tolerated. An early US secretary of the treasury, Albert Gallatin, said in 1807 that "government tends to be incompetent" (Rosenkranz, 2015). Note the singular: Gallatin is referring here to government in general—the human institution of governance—not to particular administrations or political systems. His considerable experience (nobody has served longer as secretary of the treasury) was with a young democratic republic, but other forms of government, both historical and modern, display similar tendencies toward breathtaking ineptitude. Thus, incompetent government by the democratic monarchy of Great Britain led directly to the birth of the young republic which Gallatin served. Churchill once famously summarized different methods of governing a nation in a House of Commons speech (November 11, 1947): "Democracy is the worst form of government, except for all those other forms that have been tried from time to time." Today, undemocratic nations with limited freedom of speech display their ineptnesses less publicly but equally obviously: China is undemocratic with little freedom of speech; it is inefficient and corrupt. Russia, even under Putin, retains some shreds of democracy and freedom of speech, and is inefficient and corrupt. By all means let us be appalled by such incompetence and let us try to root it out in our own countries, but let us not be surprised by it, for it is the norm—it is human. Peter expresses the sentiment humorously: "Sometimes I wonder if the world is being run by smart people who are putting us on or by imbeciles who really mean it" (Peter, 1969).

Modern private-sector hierarchical incompetence is expressed very well in the popular comic strip *Dilbert*. Creator Scott Adams formed the Dilbert principle, which states that a company systematically promotes its least competent employees to management in order to minimize the amount of damage they are capable of doing. Hmm, humorous but unlikely, I hope. Yet many managers (in the engineering companies that I have worked for) do jobs for which they have received no formal training and display—to me, at least—no talent for the positions in which they find themselves. A better-known formulation of human organizational ineptitude is Murphy's law: anything that can go wrong, will. This aphorism seems to date from the 1950s and is of uncertain origin; its significance lies in the fact that it is so widely known and has lasted so long. Two US Army acronyms, SNAFU and FUBAR, express similar expectations more vulgarly (the F and U stand for the same words in both).

Yet another statement of organizational ineptitude is the *law of unintended consequences*, traced by some back to Adam Smith. There are many examples from history, and particularly from environmental actions, which demonstrate the harm that stupidity, ignorance, and short-sightedness can do. Infamously, the introduction of cane toads to northern Australia in 1935 to suppress beetles that damaged cane fields was unsuccessful for two reasons: the cane toads did not suppress the beetles and became major pests in their own right. There is a long and sad procession of such examples involving introduced species.

The most elegant example of the law of unintended consequences, to my mind, is the *cobra effect*, for which an action produces an effect exactly opposite to that intended. In Delhi, India, during the time of the Raj, British rulers wanted to reduce the number of venomous cobra snakes, which were perceived to be a threat to the population. So they put a bounty on cobras; a monetary reward was given to citizens of Delhi for each dead cobra they turned in to the city authorities. So local people bred cobras, to boost their income. Then the British authorities eliminated the bounty, not wanting to encourage this behavior. So the unwanted cobras were released. End result: a well-meaning but ill-thought-out action worsened the problem it was invoked to solve. Move from Delhi to Hanoi, change cobras into rats, and British colonial authority into French colonial authority, and you find exactly the same thing. The cobra effect is not limited to colonial governments and the historical past, of course; many people see actions from the War on Terror in Iraq, for example, as exacerbating the problem that was meant to be solved.

It may seem simplistic and perhaps churlish to harp about the human inability to organize well—like focusing on a child's spelling mistakes—but there is a serious point lurking beneath the banality and humor. Fixing the global warming problem we have created will require much organization, cooperation, and coordination across the globe; we may get only one chance at it, depending on the breaks, so we cannot afford to let the Peter principle and Murphy's law create a FUBAR future due to the cobra effect.*

40 Collective Stupidity

Of course, it is easy to point to examples of human collective stupidity; nevertheless, it is fun and instructive to do so, so here are a few examples. If you disagree with me about our stupidity, then I'm sure you can replace my examples with examples of your own, and ponder over why we disagree. Thus, there are people who still believe that the Earth is flat—they have formed a society, which you are welcome to join.† There are people who do not believe that Darwinian evolution is close to a true description of biological evolution, who prefer a creationist, intelligent design origin for humanity, despite a huge corpus of evidence and the views of an overwhelming majority of biological scientists who subscribe to Darwin's big idea. There are

*See Dubner (2012), Glaser (2015), Peter (1969), and Rozenkranz (2015). The web contains many examples, gleefully reported, of the various social laws summarized here—Wikipedia is a good place to start. For the radar engineering instances cited, see Brookner (1977).

†The Flat Earth Society associate membership is free; full membership (which gets you a signed certificate) requires a small donation. See website for details: http://www.theflatearthsociety.org/home/.

groups of people who are prepared to blow themselves up in order to kill other people who do not share their religious beliefs or, more accurately, do not belong to their group. There are groups who advocate public safety by arming rather than disarming society. There is the insane rationality of Mutually Assured Destruction, a nuclear standoff between potential enemies that has avoided a global nuclear war, so far. (Yes, MAD has worked, but why did it have to be invoked in the first place?)

All these examples are collective—of groups rather than of individuals. Individual stupidity might be watching reality TV, or voting for Donald Trump against your own interests, or sniffing glue, or believing that smart grids cause cancer or that the Beatles were a communist plot. Collective stupidity is more dangerous to human well-being, in general, because it involves more people—more people exhibit it and are affected by its consequences. Stupidity itself is not necessarily worrisome—as an individual trait it can be endearing or amusing. Stupidity is not correlated with intelligence or lack of it; for example, we all know clever people who are also in some ways really stupid (geeks, for example). Stupidity has something to do with irrationality: a stupid person is irrational, but an irrational person is not necessarily stupid. I love my wife—this is not rational, nor is it stupid. Stupidity often goes together with hypocrisy ("Thou shalt not kill, except for THEM"), though I doubt I could make a watertight case that they have to be connected.

In short, stupidity is human, perhaps uniquely so. A wild animal with just enough brain to survive has no leg room for stupidity. A stupid snake, an asinine ass, a dumb donkey, or a silly spider would go extinct before we got to know about it. By developing complex societies, we humans have given ourselves some wiggle room so that a below-par specimen can survive—a beaver that is blind would not last long in the wild, but a blind human can live a long and fulfilling life. Stupidity gets a similar break; it is not bred out of humans by the cold relentless force of natural selection, because we are protected by social rules that raise us above our survival-of-the-fittest origins.

For individuals, that is. Collective stupidity has the potential to be harmful or even fatal for the species.

•   •   •

Crowds behave differently from the people who compose them. The psychology of crowds, or mobs, or social movements—their collective behavior—generally violates social norms; their group dynamics encourage acts that normally would be unthinkable.* One of the five theories that psychologists and sociologists have put forward to explain the phenomenon suggests that a driver for this excessive behavior is loss of personal responsibility, for an individual who is part of the crowd. There is also a well-known *bandwagon effect*, in which a fad (say, wearing a baseball cap back to front) or belief (a religion) or rumor (smart grids cause cancer) is more likely to be taken up by an individual within the crowd if she sees others around her adopting it. This is positive feedback, and results in exponential growth of the fad or belief or rumor. The bandwagon effect is a consequence of the desire of many people to conform—such folk are sometimes derided as *sheeple*. The effect has been well established by experiments in psychology.†

A collective intelligence also exists among groups, in which it is clear that the whole is greater than the sum of its parts. For example, within a research community, the individuals can "spark" off each other and perform better than the same people would in isolation. It is perhaps not difficult to see how such a collective intelligence arises: the knowledge of the group is greater than that of each individual, and might be communicated to each member of the group, spurring progress. But there is another side to this coin, which, unfortunately, often dominates within institutions. Chomsky calls it "institutional stupidity"; I will generalize here to any large collection of people and call it collective stupidity. Chomsky provides two examples; the first is the near-suicidal actions of nuclear weapons saber rattlers during the Cold War (which includes the MAD strategy, to which I have already alluded), and the second is the priority list of today's CEOs—much

---

*I recall once seeing, in Paris, a visiting group of Scottish soccer fans (I suppose "clan" would be the appropriate collective noun) forming a line across a city-center street and collectively raising their kilts, to the amusement of passersby and the local police.

†I refer here to the Asch conformity experiments of the 1950s, which shed light on the psychology of groups. The behavior of crowds and of institutions, and the difference between this behavior and that of individuals, is discussed by, e.g., Locher (2001), Kahler (1998), and Revkin (2012).

more mundane but also more common. An annual study by Pricewater-houseCoopers, a prominent professional services company, places climate change as the *twentieth* most important priority of CEOs, or lower. It is not difficult to guess where this attitude comes from: according climate change a higher priority might cost a business quite a lot of money, and so would depress short-term profits. If short-term gain is indeed the explanation, then this blinkered view counts as an example of collective stupidity, given that most intelligent people (I think we can presume that CEOs are intelligent) consider the subject much more important and urgent. It would be interesting to learn if each CEO places climate change that low on his or her personal priority list.

Beyond business institutions, there are plenty of examples of collective stupidity between nations. Often irrational, or rationally exploiting a popular but irrational mood, the practice of international relations is far from the naive ideal—a coming together of different countries to find an optimum strategy that provides maximum benefit for the world as a whole. Thus "national honor" may be invoked by one country to justify attacking another. This is irrational, and produces no benefit at some cost; what does "national honor" actually mean? If one country—say France—attacks another to obtain revenge for some act, at least that motive has elements of logic to it—it placates a provoked populace and discourages future acts against France. But "national honor"? It was invoked by Russia to justify the occupation of Crimea, along with another stupidly irrational concept: "sacred land."

Dictators bring individual stupidity to the international stage and thus their actions, because they influence nations and affect the collective behavior of humanity, can fall under the heading of collective stupidity. It is hard to justify, even from the narrow perspective of benefiting the perpetrators, some of the recent actions of certain terrorist groups, such as IS, or of the North Korean government. It is also depressingly easy to interpret the decision to go to war with Iraq in 2003 as a clear example of collective stupidity. Several intelligence agencies decided wrongly that Saddam possessed a WMD capability—perhaps we can infer a bandwagon effect at work within these agencies—and they convinced the US and British governments, among others, to invade that stricken country. Interestingly and significantly, at the time, a majority or large minority of the populations of the United States and Britain, and of other democracies, vociferously opposed this action.

I do not know if it is possible in practice for a numerous and dominant species of inventive, clannish, aggressive, loving, and self-ish individuals to act collectively with more intelligence than they act individually. Logically it must be possible in principle, but history shows over and over that it appears not to be the case for the species to which I belong. For reasons that I suppose must be in our DNA, we much more often display collective stupidity than collective intelligence, though we are capable of both. In the Anthropocene there will be great and urgent dangers posed to us by climate change, antibiotic apocalypse, changing food supply, and population growth. These dangers will be exacerbated by our collective stupidity and eased by, perhaps only by, any collective intelligence that we can muster and bring to bear on the issues.

•   •   •

The last few sections have pointed out how we are a stupid species—the precise manner of our stupidity and incompetence. Yet, of course, this is not the whole story because, quite simply, if it were, Darwinian natural selection would have seen us go extinct before we ever got to the stage of producing a Darwin, or of reading a book, or of developing a technology powerful enough to get us into trouble. The other side of the coin is our cleverness, the inventiveness that gave rise to the industrial revolutions and the technologies of today that make our lives better than those of our forebears who invented them. So how will those two opposites gel, to determine (influence, rather, since the world is not deterministic) our future?

 **ABC but Not D**

Let me summarize the glum trifecta of human frailties—argued earlier in chapters **33, 35,** and **40**—that in my opinion lower the bar of reasonable expectations for how we are going to cope with climate change, overpopulation, and possible shortfalls in food supply and effective antibiotics in the decades to come:

(a) Nobody understands economics
(b) Disparate groups cannot act cooperatively for the common good
(c) Collectively, we act more stupidly than we do individually

By definition, when we enter the Anthropocene epoch we will be influencing the surface of our Earth mightily—we already are, in climatological if not yet stratigraphical ways. Thus climate change depends on economic health (which will dictate industrial output, including carbon dioxide), which is unpredictable. So we cannot place too much faith in long-term climate predictions, because we can't predict long-term economics. This lack of confidence in climate predictions leaves wiggle room for climate change skeptics and other short-termists who adopt a Prisoner's Dilemma attitude of selfish gain; as a species we will find it difficult, as we always have, to act in a concerted manner for the common good. So don't expect too much from climate talks—whatever is said and agreed on, our response will be slow and inadequate in practice. Thus we *will* exceed the 2°C guardrail and may well go quite a bit beyond it. We know this is stupid, but we just can't get our act together to do as well as we might.

• • •

How well would it be possible to do, if we were not human but instead were some sort of goody-goody species of altruistic self-sacrificing paragons of virtue capable of cooperative action? (I'll call these insufferable creatures Tweety-Pies; they are humans without the a, b, and c listed above.) Let's not go down the road of saying that, if we were Tweety-Pies, we would not be in the mess we currently find ourselves, because that would be unhelpful—we must start from where we're at. So let's agree that, for the next several paragraphs, 7.4 billion Tweety-Pies inherit our Earth; they occupy it in our place and are faced with a climate that will soon be changing in alarming ways. What would they do, and what difference would it make?

Emission reductions and carbon taxes would be implemented worldwide to reduce anthropogenic carbon dioxide effluvia. Fossil fuels would be phased out over the next decade after a worldwide building program of several thousand nuclear power plants brings this stopgap source of power online, until renewables can take up the slack. Needless to say, this building program will include long-term storage sites for radioactive spent fuel. After a generation or so, the nuclear plants could be reduced in number (as renewable power sources kick in), though not eliminated entirely because, as we have seen, renewables won't be able to provide us with all our power needs. Thus a mixture of nuclear fission and renewable power sources (hydro, wind, solar) will see the Tweety-Pies through to the twenty-second century when, I hope, nuclear fusion reactor technology matures to the extent that power from this clean and abundant source becomes economically viable. Thereafter, the provision of safe and sustainable power will cease to be a serious issue, whatever the global power requirements become.

Paying for all this development will strain economies, and I won't make any predictions about that. The high-consumption lifestyle that some of us humans have enjoyed over the last couple of generations will take a hit, but the Tweety-Pies won't mind because they know that it is all in a good cause, and that works for them. The burgeoning global population is the ultimate cause of all the stresses that will have to be faced, because increased numbers of people have led to increased power consumption, more pollution and atmospheric carbon, more food consumption and increased diseases (and antibiotic-resistant bacteria). Stepped-up investments will be required to accelerate a second green revolution, and to discover new antibiotics. Programs of education, economic aid, and perhaps also voluntary migration and one-child policies will result in a global total population that is at the low end of the predictions for humans, perhaps "only" nine billion by the end of this century. The relatively low number of Tweety-Pies will reduce the impact of the other stress factors, compared to what we humans will have to face.

Despite all these efforts and the self-sacrifice of the hardworking Tweety-Pies, they will still suffer bad effects of climate change, because they inherited from us humans an atmosphere that has been pumped up with GHGs. They will face changing weather patterns with increased droughts in dry areas, increased flooding in wet areas, and more extreme weather of all kinds happening more and more often, with differential effects in different parts of the world. The differences—with poor and developing nations suffering more than rich and developed ones, on average, because of latitude—will be lessened, because the Tweety-Pies will manage immigration and aid programs better than we will. They will suffer less anyway, except economically, because they will implement expensive but effective technofixes that will sequester atmospheric carbon and reduce carbon dioxide levels at a rate faster than would occur naturally. For this reason, while there will be an overshoot of global average temperature rise beyond the guardrail level of 2°C, it will be temporary for them. Temperature increases will fall below the guardrail level by the end of this century or during the next.

With these Herculean efforts and with a great deal of self-sacrifice, the Tweety-Pies will pay for our mistakes and will right the ship that we have sailed into dangerous waters. Thereafter, they will maintain a steady population, resume economic growth and prosperity (on average, subject to the usual weasel-words about inherent unpredictability), grow industry and commerce with clean power, and look back at us from the distant future, wagging their fingers. Their world will suffer from climate change effects of decreasing intensity—it will be like the rough seas that follow a storm. The residual effects (which may last centuries) will be due to the long timescales on which the dynamics of our climate operate.

●   ●   ●

So much for the idealized version of what COULD happen. This scenario is not realistic because we are not Tweety-Pies, but it has been a useful exercise to go through, because it shows that we are not yet doomed (in my opinion; some experts disagree, as we have seen). In the next and last section of this book, I get to speculate about what I think is a realistic possibility for the long term future of our planet with real humans (not Tweety-Pies) in charge of it. Here I will concentrate on explaining why I think we are not (d) doomed—despite the a, b, and c listed earlier in this chapter.

We saw that technofixes are possible in theory though uncertain in practice—uncertain because they will involve more cooperation than I believe we are capable of, and because we do not know the end result of applying these fixes. Such fixes include massive schemes to brighten clouds and spread aerosol particles into the upper atmosphere, to lower albedo and thus reduce the Earth's solar energy intake. Ideas such as these may be a brilliant silver (iodide; used in cloud seeding) bullet or they may be a disastrous mistake; many people will argue that, given our uncertainty, it would be irresponsible to proceed with such mitigation measures without significantly more insight as to their downstream consequences.

Nevertheless, some approaches to reducing the carbon content of the atmosphere can be put in place without much risk. Thus, we can aim to decarbonize the world economy by planting trees and building artificial trees that will take carbon out of the atmosphere (carbon capture and sequestration–CCS). This process will be slow—results may not show until the next century—but it is something that will help. A little less slow and still safe: we can take a number of steps to filter out soot that humans generate from various processes, with clean-burning stoves, particle filters on vehicles, bans on burning agricultural waste, etc. NASA estimates that by doing this and taking other equally simple measures* we can, over the next 40 years, eliminate half the anthropogenic global warming that we would otherwise produce. Of course, this is slow and expensive and some people/industries/countries will cheat, so the effectiveness of this type of mitigation will be compromised to a greater or lesser degree, depending on how much of the world can be signed up to such a program. The simpler ozone hole problem looks like it is on its way to being solved, and we can see a path forward to at least reducing the deforestation problem, in part. So, some degree of forward thinking and of collective action by our species is possible, and consequently I think it is reasonable to expect that humans will be able to implement a measure of global warming mitigation, though not the whole Tweety-Pie program. We will not see a Business as Usual scenario unfolding in future decades, thank goodness.

As well as *mitigation*, we can aim for *adaptation* to the coming climate changes. We can anticipate from climate models the type, if not the degree, of climate changes that can be expected in each country of the world, so each country can take its own steps to prepare its pop-

*Other measures to reduce methane production and ozone production. We have seen that methane is a potent GHG. Ozone is desirable in the stratosphere but not in the lower troposphere. See the NASA webpage "Responding to Climate Change" and links within it. http://climate.nasa.gov/solutions/adaptation-mitigation/.

ulation for what is about to hit them. Adaptation is more likely to be implemented than mitigation because it entails reacting to a climate disaster that has already happened or implementing preventative measures for climate events that are deemed imminent. Governments have no difficulty in selling such adaptation policies; indeed, they are often under pressure from their public to do more. There is little opposition from vested interests (in marked contrast to, for example, the foot dragging of fossil fuel industries to climate change mitigation strategies such as CCS). So we can confidently anticipate that communities really will adapt to the perceived climate changes that are heading toward them, according to their means.

According to their means—here is another example of rising inequality. Of course, rich nations can take more steps to avert the bad effects of coming climate change than can poor nations and so will suffer less. They will suffer less anyway because, as we have seen, they are geographically favored to win rather than lose from climate change, on the whole. Many but not all adaptation techniques are expensive, such as building (or raising or strengthening) seawalls and levees. In some areas, desalination plants will be needed to counter reduced drinking water levels. Less expensively, improved irrigation and rainwater storage can reduce water losses. Improved city drainage and modified infrastructure can help prevent flooding or reduce its effects. Thus, Singapore has built its mass transit system at least one meter above the highest recorded flood level. Here is one example of what has been described as a champagne result on a beer budget: when a city road needs fixing, it can be recrowned a couple of inches higher to divert water during heavy rainfall. Permeable pavement can be installed to aid water runoff. Every little bit helps.

Adaptation to changing average temperatures will reduce anticipated losses in food production: we can adjust planting dates and crop varieties; relocate crops and livestock; and develop strains of crops that better tolerate heat or drought or flooding. Affordable measures to reduce the effects of increasing temperatures include installing air conditioning in schools and planting heat-tolerant trees to line streets. Climate change will redistribute diseases (for example, malaria, dengue, and other tropical fevers will spread northward). Surveillance and control of diseases and the development of health action plans in the event of outbreaks of these new (to some regions) threats to public health will be necessary adaptations. Infrastructure protection is already anticipated and is a measure of adaptation that will be demanded by the public based on past failings during times of disaster. Power plants will be made secure from flooding, power transmission and distribution lines will be strengthened, second sources of power will be available through improved (smart) grids. Cellphone networks will be made more robust.

All of these adaptations are, in varying degrees, already under way in different countries according to their wealth and in light of past experiences with extreme weather events that are, rightly or wrongly, attributed to climate change. Thus in the United States, the Chicago heat wave of 1995, Hurricane Katrina, and Superstorm Sandy have all left their marks, and the communities that suffered are taking steps to reduce the effects of future heat waves, hurricanes, and superstorms. Adaptations will be costly, though not as costly as geoengineering mitigation, and not nearly as expensive as doing nothing (according to economic studies, for what they are worth). Adaptation may be (has been seen by some critics as) an admission that mitigation is not going to work, but in fact adaptation is a result of simply being realistic. Adaptation has a good chance of being adopted to a considerable degree and for many decades into the future, and it will lessen the effects of global warming on human populations. So here is one reason for justifying a less pessimistic view of the future than that of, for example, James Hansen (who foresees, among other dangerous climate developments, sea-level rises of several meters in the next couple of centuries, which would indeed be catastrophic for humankind).*

A rosy view of our future is as silly and unrealistic as my name for the creatures who might implement it—Tweety-Pies. A truly pessimistic view (that we are toast whatever we do now, because an irrevocable chain of events that we have put into place is about to unfold) is premature, in the opinion of most scientists. Our nature precludes decisive collaborative action to mitigate the full effects of global warming; our technology and self-interest can alleviate many of the short-

---

*To read more on climate adaptations, see Helm (2015), Klinenberg (2013), and the NASA website. For Hansen's view on recent dangerous developments in climate change, see Hansen et al. (2015) and Mooney (2015b).

term consequences, at least in developed countries. Let us hope, but not only hope, that these adaptations will be sufficient.

• • •

So that's it—the kind of future that we will likely be bequeathing. My account of the near future ends the book, except for some more-whimsical ruminations about how the Anthropocene might unfold in the longer term, for a few of the people who will be living in it.

㊷ Where Are You Going, My Little One?*

A few hundred years in the history of a small blue-green planet may not matter much in the universal scheme of things, but this place called Earth interests me (because, as the bumper sticker says, it's where I keep all my stuff) and so here goes with my vision—hallucination may be a better word—for our mid-term future, say up til 2200. A word of caution first expressed in chapter **30** is perhaps worth repeating here: every long-term prediction is very speculative. To emphasize this point (the unpredictability and contingency of future events), and to leaven the text, in this final chapter I embed a few biographical sketches and snapshots of people yet to be born within a larger body of future history that will outline the road ahead.

That was my plan. Then it occurred to me that this "larger body of future history" will be impossible to write with any credibility. So, at the outset, I ditch this body of work entirely—strangle it at birth—leaving you only the broad brushstrokes, the insightful if fanciful outlines of a few citizens of the Anthropocene.

• • •

On April 5, 2025, Liliane Algafari gives birth to a daughter, Yana, in the German city of Dusseldorf. Liliane and her husband Mohammad

* Lyric from song "Turn Around," written by Malvina Reynolds, Alan Greene, and Harry Belafonte in 1963.

had emigrated from Syria a few years earlier. A smart kid who develops an interest in biochemistry, Yana attends Hannover Medical School and embarks on a career in research, funded generously by the Chan Zuckerberg Initiative. At the age of thirty-seven she discovers a whole new class of antibiotics, developed from a parasitic mite found only in the eyelids of a rare type of Ecuadorian tree frog. The year in which the number of effective antibiotics reaches a maximum, prior to the work of Yana and her team, is 2031, and the three decades following "peak antibiotics" see many epidemics of once-curable diseases decimate populations, mostly in developing and underdeveloped countries. In 2066 one of Yana's new antibiotics is fast-tracked through tests and distributed widely, proving very effective in eradicating a screaming hab-dab pandemic that is decimating the populations of the Gobi Desert, the American Southwest, central Iran, Spain, Australia, and other elevated dry areas of the world. (The prominent symptom of this terrible disease is a dry cough; a prominent victim of this particular pandemic is Professor Albedo of the University of Phoenix.) For her pioneering work, Yana Algafari is awarded the 2080 Nobel Prize for Medicine.

● ● ●

Barsha Lal is thirty-nine years old when, in July 2050, she is appointed by the Indian government to negotiate a dispute with Bangladesh over the division of water from the shared Ganges River. Though these negotiations fail and the dispute escalates, both sides are impressed by Ms. Lal's skill as a negotiator. She is nominated for a position at the United Nations, which she takes up in 2052. That year, the number of people in the world who do not have access to safe drinking water exceeds two billion for the first time. Lal is the chief UN negotiator mediating the Nile River Basin crisis of 2055, the second and more serious of two water wars in this region. Skirmishes along the common borders of Egypt, Sudan, and Ethiopia flare into war following the breakdown of talks over irrigation rights. Three years later, Lal is more successful in mediating between Turkey and the Arab States— formerly Syria and Iraq—over Turkish damming of its eastern rivers, including the Tigris and Euphrates, that flow southeastward through the Arab States. In 2059 she is again in the Middle East negotiating a peace between Israel, Lebanon, Palestine, and the Arab States over increasingly scarce water resources. The peace is short lived.

Tragically, Barsha Lal dies in a freak flood in Tajikistan, in December 2071, negotiating a peace between that country and neighboring Uzbekistan; the two countries were warring over the Amu Darya River tributaries.

* * *

The shell of the brewery main building is old, and is maintained in the style of a century and a half earlier when it had been built, but in the year 2081 everything else about the building and the industry it represents is different. The names on Bud's office door still include B. Weiser, he is pleased to see, plus half a dozen other names of technicians with whom he timeshares his job in the production department. Very few people actually work on site these days; beer production in Milwaukee has changed a lot since the days of Millercoorsbudpabst. The enormous Chinese brewing conglomerate Sù Zuì takes over and modernizes the plant, doubling production and halving the workforce yet again. Apart from the reduced workforce, all that is retained from the old times is the beer can label.

Bud is called in to see if a partial reboot is necessary after the latest lightning storm. Supercell thunderstorms are now quite common in Wisconsin—they have spread east from the Great Plains in recent decades, and they can wreak havoc with electrical systems. Bud thinks much of the Midwest is heading east these days—he can see miles and miles of aging wind turbines when he drives outside the city. The brewery has upgraded to the new smart grid (and is in the throes of constructing infrastructure for the much-anticipated Supergrid), so here at least the storms should not have too bad an effect. The old smart grid could cope with blackouts—they were a thing of the past for nearly everyone. The American part of the Supergrid is only partially built and is not yet up to speed, so Bud needs to check the quality of the electricity supplied by the generators. Sometimes, when they cut in, a few of the production line robots cut out. His interface terminal is filled with messages from the machines telling him what they think needs looking at (a lot) and what they think of him (not much).

Bud is not so concerned about the threat to his brewery job during the current recession, because his other business is going quite well. He sells electricity to the city; it comes from V2G when his car is in the garage, from the solar panels on his roof at home in the 'burbs, and from panels on land that he rents from a neighbor. Bud is also contracted to sell power to a retirement community in Portland, Oregon; he takes advantage of the time difference to provide solar electricity that is generated during Milwaukee summer mornings to heat Portland water before the sun rises there. He is thinking of cofounding, with his geek sister Gertie, an electricity arbitrage start-up that will be easily accessible to the public from an app. If it works, he might be able to afford a vacation. Maybe he will call the startup "Dial-A-Volt," though that has an old-fashioned ring to it. "Power Up" is much better.

• • •

Giovanni Diaspro begins his working career in the fall of 2095, after graduating with a technical degree from the University of Naples. He is employed as a technician in the carbon sequestration industry in southern Italy for twenty years, working mostly with genetically modified trees. He marries Antonella in 2102 after being promoted, and starts a family. He considers emigrating during this period; the Mezzogiorno is changing, and he fears that his children will not enjoy the quality of life that he and his wife enjoyed when they were youngsters. Immigration from Africa has changed the face of the region, increasing population density and social strife, and placing pressure on the local social services. This is because the better educated and more employable immigrants have tended to move on, northward to Milan and Turin, where the good jobs are; those remaining tend to be in a more desperate situation and have not integrated well, creating social tensions. Also, the heat and the Mediterranean storms seem to be worse and more frequent than those he remembers from childhood.

The Diaspros move north to Rome and lodge with Antonella's parents for a while. Giovanni is able to obtain only temporary jobs, working for the Climate Adaptation Department of the civil service. When these jobs dry up, he applies abroad and is able to find work in a Swedish vineyard. In May 2110 Giovanni, Antonella, and their two young children move to Småland, to one of the better-known young vineyards that have put Swedish reds on the vintners' map of the world. Giovanni's position is more junior than his job in the Mezzogiorno, and is not well paid, but he and Antonella like the area and their kids' school, so they settle. They miss their families, but are cheered by the growing Italian community in their district.

• • •

"Peak food" is the year 2073, when Podraig O'Raifeartaigh is born. Many millions of people suffer from malnutrition and die in the three decades between this time and 2103, the year O'Raifeartaigh is awarded a Bill and Melinda Gates Research Scholarship at the Filipino Research Institute for Engineered Species (FRIES). His work at FRIES leads to the development of genetically engineered strains of potatoes that are salt, heat, drought, or flood tolerant. They retain the characteristically high energy density of natural potatoes (among the highest recoverable food calories per acre of any plant). The engineered potatoes of O'Raifeartaigh's team are distributed to selected test sites in South Asia, Central Africa, and Central America in 2109. By 2112 it is apparent that the new strains are vigorous. Further work develops variants that are tolerant to the latest pesticides and fungicides, yet are safe to eat. By the year 2125 these potatoes are planted and grow wherever humans live, providing nutrition where previously there was malnutrition. O'Raifeartaigh and his team at FRIES are awarded the 2133 Nobel Peace Prize.

• • •

Nigel Deuteron is born in London on May 16, 2009. His father, Torr, is a professor of physics and his mother, Pascale, is a civil engineer. Young Nigel escapes the pressures of his domestic life by reading books on nuclear physics, and it is no surprise to his parents when he chooses a career in this field. After a stellar university career, in 2110 Dr. Nigel Deuteron enters Glasgow University and 10 years later establishes a fusion power research team at Costa del Clyde—a resort village in the west of Scotland more used to seeing wealthy Italian tourists escaping the insufferable heat of a Mediterranean summer or well-heeled Greenlanders escaping their dark winters.

Key advances in fusion power research had been made steadily if slowly over the previous century. Until the 2030s, the fusion community explored many different types of fusion reactor designs; it was not until 2040 that the best approach was finalized. Another 20 years are required to produce a reactor that is stable and produces more power than it consumes. For several decades thereafter, fusion research slows, because of funding cuts during a period of widespread economic recession—fusion research is very expensive. The vital contribution of Deuteron's group, which was partially funded by the Weiser Foundation (established by a brewery engineer who made a fortune in electrical power arbitrage a generation earlier), was to develop a process of design iteration whereby the cost of generating electrical power from a fusion source was cut and cut until it was able to compete with the price of electricity from solar and nuclear fission. This work consumes Deuteron's entire career and those of his (mostly Chinese, American, and French) coworkers. Progress is slowed further by the very success of other renewable power technologies and of fourth-generation fission reactors, which reduce power costs. Deuteron dies in his ninety-eighth year and is acknowledged as a key figure in bringing fusion power to the world. The first generation of commercial fusion reactors is built during the last decade of the twenty-first century; thereafter the generation of power to meet the ever-growing demands of humankind ceases to be a major issue, for the first time in 450 years.

● ● ●

Buford Beauf is turned in to the authorities after a tip-off from his washing machine during the evening of February 7, 2130. The appliance earns a Good Citizenship Award from the civic community in Iota, Louisiana, where Beauf is living at the time of his arrest. He had moved to the new state capital nine years earlier, after Baton Rouge was abandoned following the Great Storm of 2121. Beauf is tried for the remote fly-by murder of 11 people who were attempting to enter the United States illegally via Mexico, and is convicted. His appeal is supported by the Armed Wingnuts who, since formally separating from the Republican Party at the end of the previous century, have often championed gun-rights issues. Despite their support, Beauf's appeal is lost. This event marks the first major success in challenging the second amendment to the United States constitution in its 339-year history.

The case turns on the nature of the crime. Beauf prints a copy of a "Saturday night special" hunter-killer drone of the kind commonly sold in hardware stores, and directs it from his living room in Iota.

The drone flies over a section of the e-wall along the border between Texas and Nueva León state in Mexico and intercepts a group of 30 migrants from Central America who are attempting a border crossing in broad daylight. This section of the e-wall is monitored and patrolled by a private company subcontracted by the US Army, which oversees the southern border. Beauf's victims had been outside the kill box—they were on the Mexican side of the border—at the time they were intercepted and destroyed. This is the reason why Beauf's appeal is rejected, despite the "hot pursuit" clause in the US-Mexico agreement on border security.

A year before the crime takes place Beauf's ex-girlfriend, Lieutenant Dixie Roan, is promoted from a low-ranking position in a Texan border patrol station and is sent to an elite Air Cavalry unit in White Sands, New Mexico, where she remotely operates large interceptor drones along one of the border hotspots that constitute the theater of Operation Stonewall, the Army's most important deployment at this time. Beauf's career heads in the opposite direction after he is dishonorably discharged for moonlighting as a washing machine repairman and evicted from the same border patrol station. Testimony from Roan and from ex-colleagues, friends, household appliances, and cars chronicle his hatred of immigrants.

After many legal appeals, Beauf's conviction for murder is upheld, and he is executed in 2169. Roan's drones intercept and neutralize numerous illegals, and she retires from a very successful military career in 2152. In retirement she writes the definitive technical history of washing machines.

. . .

During the early morning of July 1, 2159, the population of Canada exceeds for the first time that of its southern neighbor, at 536,870,912. In fact, the unheralded event marking the exact transition occurs at 02:29 that morning, when a tired Fred Sandoval and his family cross the border in their beat-up driverless electric truck. Fred emigrates from the United States to Canada in the hopes of improving the lives of himself and his family; they drive up from Texas, where he had worked shifts in the security industry and other service sector jobs in a dying state that is unable to import enough water to maintain its population and is imploding with social problems. Fred is hoping to get a job as a gofer in the Brollywood movie industry centered in Vancouver, Canada's largest metropolis. This doesn't work out, so the Sandovals drift to sunny Churchill, Manitoba, like so many others before them, and Fred gets a job at the port, where he works until he retires at the age of 80. A lifelong baseball fan, Fred becomes an avid supporter of the Churchill Moose, a major league franchise that wins the World Series three times in the first decade of the 2200s.

Bibliography

The sources used for the writing of this book are dense in number—approximately four sources for every thousand words—though for the most part the written material is not dense reading. They are a mixture of secondary popular-level articles for readability and primary technical papers for depth.

Achenbach, J. 2015. "Short-term fixes for long-term climate problems? Not so fast, experts say." *Washington Post* (February 9).

Achenbach, J. 2010. "Electrical grid." *National Geographic* (July).

Adams, G. 2011. "Lost at sea: On the trail of Moby-Duck." *Independent* (February 27).

Adger, W. N., and J. Paavola. 2006. *Fairness in Adaptation to Climate Change* (MIT Press).

Aitkenhead, D. 2008. "James Lovelock: 'Enjoy life while you can: In 20 years global warming will hit the fan.'" *Guardian* (March 1).

Alexandratos, N., and J. Bruinsma. 2012. "World Agriculture towards 2030/2050." Food and Agriculture Organization report (June), available online.

Amos, J. 2016. "'Case is made' for Anthropocene Epoch." BBC News (January 8).

Amos, J. 2015. "Earth's trees number 'three trillion.'" BBC News (September 3).

Amos, J. 2006. "Deep ice tells long story." BBC News (September 4).

Anderson, C. 2014. *Makers: The New Industrial Revolution* (Crown Business).

Anderson, C. 2012. "The internet has created a new industrial revolution." *Guardian* (September 18).

Anderson, R. 2015. "Nuclear power: Energy for the future or relic of the past?" BBC News (February 16).

Ansolabehere, S., et al. 2007. *The Future of Coal* (MIT Press).

Ashton, D. 2013. "Mining Your iPhone." Infographic on Business 2 Community website.

Audubon, J. 1827. "On the Passenger Pigeon." In *The Birds of America*.

Autin, W. J., and J. M. Holbrook. 2012. "Is the Anthropocene an issue of stratigraphy or pop culture?" *GSA Today* 22, 60–61.

Avery, M. 2014. *A Message from Martha: The Extinction of the Passenger Pigeon and Its Relevance Today* (Bloomsbury USA).

Babbin, J. 2015. "Where does nuclear power fit in our future?" *Washington Examiner* (May 18).

Baker, A. 2015. "How climate change is behind the surge of migrants to Europe." *Time* (September 7).

Barr, C., and S. Malik. 2016. "Revealed: The 30-year economic betrayal dragging down Generation Y's income." *Guardian* (March 7).

Basu, S. K., et al. 2010. "Is genetically modified crop the answer for the next green revolution?" *GM Crops* (March–April), doi: 10.4161/gmcr.1.2.11877.

Bauman, Z. 2000. *Globalization: The Human Consequences* (Columbia University Press).

BBC News. 2013. "What is fracking and why is it controversial?" (June 27). http://www.bbc.com/news/uk-14432401.

Bellassen, V., and S. Luyssaart. 2014. "Carbon sequestration: Managing forests in uncertain times." *Nature* 506, 153–155.

Benko, R. 2013. "Much bigger than the shutdown: Niall Ferguson's public flogging of Paul Krugman." *Forbes* (October 21).

Biello, D. 2014. "Ozone hole closing up, thanks to global action." *Scientific American* (September 15).

Biello, D. 2013. "How much will tar sands oil add to global warming?" *Scientific American* (January 23).

Bilton, N. 2014. "The rise of 3-D printed guns." *New York Times* (August 13).

Blackhurst, C. 2015. "It's all over for the Brics countries now—upstarts are taking the initiative." *Independent* (August 19).

Blyth, M. 2016. "Global Trumpism—Why Trump's victory was 30 years in the making and why it won't stop here." *Foreign Affairs* (November 17).

Booker, C. 2008. "Global warming: Reasons why it might not actually exist." *Telegraph* (December 30).

Borbely, A.-M., and J. F. Kreider (eds.). 2001. *Distributed Generation: The Power Paradigm for the New Millennium* (CRC Press).

Borenstein, S. 2015. "Americans largely unconcerned about climate change, survey finds." Huffington Post (November 3).

Boston Globe, editorial. 2014. "China's vow to cut emissions offers huge lift to climate talks." (November 13).

Bourne, J. K. 2007. "Biofuels." *National Geographic* (October).

Bradshaw, T. 2015. "Intel chief raises doubts over Moore's Law." *Financial Times* (July 15).

Bremmer, I. 2015. "These are the 7 challenges of globalization." *Time* (March 21).

Brennan, J. 2016. "Grim mood pervades Aspen meeting of top terror pros." *PBS Newshour* (July 30).

Bressan, D. 2011. "Climate, overpopulation and environment—the Rapa Nui debate." *Scientific American* (October 31).

Bressan, D. 2010. "The granite controversy: Neptunism VS Plutonism" *Scientific American* (October). History of Geology website.

Brink, S. 2015. "Don't take a deep breath: Outdoor pollution kills 3.3 million a year." *NPR News* (September 16).

Broderick, D. 2014. "Crowdfunding's untapped potential in emerging markets." *Forbes* (August 5).

Brookner, E. 1977. *Radar Technology* (Artech House).

Caiazzo, F. et al. 2013. "Air pollution and early deaths in the United States. Part I: Quantifying the impact of major sectors in 2005." *Atmospheric Environment* 79, 198–208.

Cardwell, D. 2015. "Offshore wind farm raises hopes of U.S. clean energy backers." *New York Times* (July 23).

Cardwell, D. 2012. "Tax credits in doubt, wind power industry is withering." *New York Times* (September 20).

Carey, J. 2012. "Is global warming happening faster than expected?" *Scientific American* 307 (November).

Carrington, D. 2016. "The Anthropocene epoch: Scientists declare dawn of human-influenced age." *Guardian* (August 29).

Carrington, D. 2014. "World population to hit 11bn in 2100 with 70% chance of continuous rise." *Guardian* (September 18).

Cassidy, J. 2012. "Ferguson vs. Krugman: Where are the real conservative intellectuals?" *New Yorker* (August 20).

Caulfield, C. 1984. "Acid rain study under fire." *New Scientist* (September), p. 3.

Ceballos, G., A. H. Ehrlich, and P. R. Ehrlich. 2015. *The Annihilation of Nature: Human Extinction of Birds and Mammals*. (Johns Hopkins University Press).

Ceballos, G., et al. 2015. "Accelerated modern human-induced species losses: Entering the sixth mass extinction." *Science Advances* (June), doi: 10.1126/sciadv.1400253.

Chase, S., and J. Jones. 2015. "Clinton's Keystone pipeline opposition puts Liberal, Conservative plans in question." *Globe and Mail* (September 23).

Chomsky, N., and L. Polk. 2013. *Nuclear War and Environmental Catastrophe* (Seven Stories Press).

Clark, D. 2015. "Moore's law is showing its age." *Wall Street Journal* (July 16).

Clark, M. 2014. "Aging US power grid blacks out more than any other developed nation." *International Business Times* (July 17).

Clark, S. 2014. "Tropical rainforests not absorbing as much carbon as expected, scientists say." *Guardian* (December 15).

Cohen, L., and D. Ramos. 2016. "Bolivia declares state of emergency due to drought, water shortage." Reuters (November 21).

Collyns, D. 2014. "Lima climate talks: EU and US at odds over legally binding emissions targets." *Guardian* (December 2).

Conan Doyle, A. 1971. *The Complete Sherlock Holmes Short Stories* (John Murray).

Cook, L. M., et al. 2012. "Selective bird predation on the peppered moth: The last experiment of Michael Majerus." *Biology Letters* (February 8), doi: 10.1098/rsbl.2011.1136.

Corcoran, D., and B. Connors. 2015. "Father says 'flying gun' drone video broke no laws." NBC Connecticut (July 22).

Corneliussen, S. J. 2015. "Attention grows for the Anthropocene, 'an argument wrapped in a word.'" *Physics Today* (March 24).

Costello, C. 2015. "America: Too fat to fight." CNN (April 21).

Crutzen, P. F. 2002. "Geology of mankind." *Nature* 415, doi: 10.1038 /415023a.

Crutzen, P. F., and E. F. Stoermer. 2000. "The 'Anthropocene.'" *IGBP Newsletter* 41, 17–18.

Cusick, D. 2015. "Solar power sees unprecedented boom in U.S." *Scientific American* (March 10).

Cuynar, G. 2006. "From fabless to chiplaco." *Globe and Mail* (April 25).

Deane, P. 1965. *The First Industrial Revolution* (Cambridge University Press).

Denny, M. 2017. *Making Sense of Weather and Climate* (Columbia University Press).

Denny, M. 2013. *Lights On! The Science of Power Generation* (Johns Hopkins University Press).

Denny, M. 2010. *Super Structures: The Science of Bridges, Buildings, Dams, and Other Feats of Engineering* (Johns Hopkins University Press).

Denny, M. 2009. *Froth! The Science of Beer* (Johns Hopkins University Press).

Denny, M. 2007a. *Blip, Ping, and Buzz: Making Sense of Radar and Sonar* (Johns Hopkins University Press).

Denny, M. 2007b. *Ingenium: Five Machines That Changed the World* (Johns Hopkins University Press).

Desai, R. M., and J. R. Vreeland. 2014. "What the new bank of BRICS is all about." *Washington Post* (July 17).

Desmond, A. 1998. *Huxley* (Penguin).

Desmond, A., and J. Moore. 1991. *Darwin* (Michael Joseph).

Diamond, J. 2005. *Collapse: How Societies Choose to Fail or Succeed* (Penguin).

Dimick, D. 2014. "As world's population booms, will its resources be enough for us?" *National Geographic* (September 21).

Dobb, E. 2013. "The new oil landscape." *National Geographic* (March).

Dowie, M. 2001. *America's Foundations: An Investigative History* (MIT Press).

Doyle, A. 2016. "World will lose two thirds of wild animals by 2020, WWF warns." *Globe and Mail* (October 27).

Doyle, A. 2007. "Tropical losers, northern winners from warming?" Reuters (April 2).

Drake, N. 2015. "Will humans survive the sixth great extinction?" *National Geographic* (June 23).

Dubner, S. 2012. "The cobra effect: Full transcript." *Freakonomics* radio podcast (October 11).

Duggan, J. 2015. "Welcome to Baoding, China's most polluted city." *Guardian* (May 22).

Dusheck, J. 2014. "No way to stop human population growth." *Science* (October 27).

Dyer, G. 2008. *Climate Wars: The Fight for Survival as the World Overheats* (Random House Canada).

Easterbrook, G. 2007. "Global warming: Who loses and who wins?" *Atlantic* (April).

Economist, editorial. 2015. "Backwards, comrades!" (September 19).

Economist, editorial. 2014a. "A bigger rice bowl." (May 10).

Economist, editorial. 2014b. "A clearing in the trees." (August 23).

Economist, editorial. 2013a. "All eyes on the sharing economy." (March 9).

Economist, editorial. 2013b. "What happened to biofuels?" (September 7).

Economist, editorial. 2013c. "The nuke that might have been." (November 11).

Economist, editorial. 2012. "The third industrial revolution." (April 21).

Economist, editorial. 2007. "Playing games with the planet." (September 27).

Edwards, B. K. 2003. *The Economics of Hydroelectric Power* (Cheltenham, UK: Edward Elgar Publishing).

Ehrlich, P. 1968. *The Population Bomb* (Sierra Club / Ballantine Books).

Ellis, M. 2013. "Early deaths from pollution in the US total 200,000 annually." *Medical News Today* (September 1).

Evars, M., O. Stampf, and G. Traufetter. 2010. "A superstorm for global warming research. Part 7: Climate change's winners and losers." *Spiegel Online* (April 1).

Fara, P. 2002. *An Entertainment for Angels—Electricity in the Enlightenment* (Icon Books).

Ferber, S. 2013. "How the internet of things changes everything." *Harvard Business Review* (May 7).

Ferguson, N. 2013. "Krugtron the Invincible, Part 3." *Huffington Post* (October 10).

Ferro, S. 2015. "Niall Ferguson takes a big swipe at Paul Krugman [updated]." *Business Insider* (May 12).

Flato, G., et al. 2013. "Evaluation of Climate Models." In *Climate Change 2013: The Physical Science Basis*. Contribution of Working Group 1 to the Fifth Assessment Report of the Intergovernmental Panel on Climate Change. Stocker, T. F., et al., eds. (Cambridge University Press).

Folger, T. 2014. "The next green revolution." *National Geographic* (October).

Foran, C. 2015. "World leaders at G7 summit back 2 degree target to fight climate change." *National Journal* (June 8).

Franke, U. 2016. "Flying IEDs: The next big threat?" War on the Rocks (October 13).

Fraunhofer Institute, https://www.ise.fraunhofer.de/content/dam/ise/de/documents/publications/studies/Photovoltaics-Report.pdf.

Freeman, S. 2015. "'Uberization of everything' is happening, but not every 'Uber' will succeed." *Huffington Post* (April 6).

Friedman, L. 2015. "Little chance to restrain global warming to 2 degrees, critic argues." *Scientific American* (May 7).

Friedman, T. L. 2008. *Hot, Flat, and Crowded* (Farrar, Straus and Giroux).

Gabbard, A. 1993. "Coal combustion: Nuclear resource or danger?" *ORNL Review* 26.

Gallagher, J. 2015. "Analysis: Antibiotic apacalypse." BBC News (November 19).

Gallucci, M. 2015. "US-India climate change talks could pave way for billions in clean energy investments but not carbon emission controls." *International Business Times* (January 21).

Gattuso, J.-P., et al. 2015. "Contrasting futures for ocean and society from different anthropogenic CO_2 emissions scenarios." *Science* 349, doi: 10.1126/science.aac4722.

Gellings, C. W. 2015. "Let's build a global power grid." *IEEE Spectrum* (July 28).

Geron, T. 2013. "Airbnb and the unstoppable rise of the share economy." *Forbes* (January 23).

Gibb, S. K. 2015. "China reduces coal use and CO_2 emissions, boosting global climate talks." *Chemical and Engineering News* 93, 22–23.

Gibbs, S. 2015a. "At the limit of Moore's Law: scientists develop molecule-sized transistors." *Guardian* (July 21).

Gibbs, S. 2015b. "Drone firing handgun appears in video." *Guardian* (July 16).

Gillies, J. 2014. "Fixing climate change may add no costs, report says." *New York Times* (September 16).

Glaser, L. 2015. "Government incompetence is the real threat to China." Cornell University (August 26).

Goering, L. 2015. "Limiting global warming to 2 degrees 'inadequate,' scientists say." Reuters (May 1).

Goldenberg, S. 2015. "Scientists urge global 'wake-up call' to deal with climate change." *Guardian* (February 10).

Goldenberg, S. 2014. "Fracking hell: what it's really like to live next to a shale gas well." *Guardian* (December).

Gollom, M. 2016. "Why trade deals like CETA have become a 'whipping boy' for anti-globalization forces." *CBC News* (October 25).

Gosden, E. 2015. "Ending onshore wind farm subsidies 'will save hundreds of millions of pounds.'" *Telegraph* (June 22).

Greenberg, A. 2014. "How 3-D printed guns evolved into serious weapons in just one year." *Wired* (May 15).

Greenfieldboyce, N. 2015. "Scientists see U.N. climate accord as a good start, but just a start." *NPR News* (December 15).

Griffith, L. G., and A. J. Grodzinsky. 2001. "Advances in biomedical engineering." *Journal of the American Medical Association* 285, 556–561.

Guemas, V., et al. 2013. "Retrospective prediction of the global warming slowdown in the past decade." *Nature Climate Change* 3, 649–653.

Hall, J. 2015. "Many experts say technology can't fix climate change." *Toronto Star* (November 9).

Hammes, T. X. 2016. "The end of globalization? The international security implications." War on the Rocks (August 2), http://warontherocks.com/2016/08/the-end-of-globalization-the-international-security-implications/.

Hammes, T. X. 2015. "3-D printing will disrupt the world in ways we can barely imagine." War on the Rocks (December 28).

Handwerk, B. 2010. "Whatever happened to the ozone hole?" *National Geographic News* (May 7).

Hanlon, M. 2013. "Is this antibiotic apocalypse?" *Telegraph* (March 12).

Hansen, J. 2010. *Storms of My Grandchildren* (Bloomsbury).

Hansen, J., et al. 2015. "Ice melt, sea level rise and superstorms: Evidence from paleoclimate data, climate modeling, and modern observations that 2°C global warming is highly dangerous." *Atmospheric Chemistry and Physics* 15, 20059–20179.

Hardin, G. 1968. "The Tragedy of the Commons." *Science* 162, 1243–1248.

Hargrove, T., and W. R. Coffman. 2006. "Breeding history." *Rice Today* 5, 34–38. http://ricetoday.irri.org/breeding-history/.

Harrabin, R. 2015. "Next two years hottest, says Met Office." BBC News (September 14).

Harrington, R. 2015. "Here's how much of the world would need to be covered in solar panels to power Earth." *Tech Insider* (September 24).

Harvey, C., and E. Newbern. 2014. "13 memories of Martha, the last passenger pigeon." *Audubon Magazine* (August 29).

Hazlitt, H. 1988. *Economics in One Lesson: The Shortest and Surest Way to Understand Basic Economics* (Crown Business).

Heaven, P. 2013. "Ben Bernanke: Nobody understands gold prices, including me." *Financial Post* (July 18).

Held, I. M., and B. J. Soden. 2000. "Water vapor feedback and global warming." *Annual Review of Energy and the Environment* 25, 441–475.

Helm, B. 2015. "Climate change's bottom line." *New York Times* (January 31).

Helm, D. 2012. *The Carbon Crunch: How We're Getting Climate Change Wrong—and How to Fix It* (Yale University Press).

Hodges, D. A., and R. C. Leachman, "Outsourcing and offshoring in the semiconductor industry." University of California–Berkeley powerpoint presentation, available at www.eecs.berkeley.edu/~hodges/Outsourcing.ppt.

Horgan, J. 2011. "Killing environmentalism to save it: Two Greens call for 'postenvironmentalism.'" *Scientific American* (December 26).

Huesemann, M., and J. Huesemann. 2011. *Techno-fix: Why Technology Won't Save Us or the Environment* (New Society Publishers).

Intergovernmental Panel on Climate Change. 2013. "Summary for Policymakers." In *Climate Change 2013: The Physical Science Basis*. Contribution of Working Group I to the Fifth Assessment Report of the Intergovernmental Panel on Climate Change. Stocker, T. F., et al., eds. (Cambridge University Press).

International Energy Agency. 2015. "Short-term energy outlook" (September 9).

Jackson, D.C. 2005. *Building the Ultimate Dam: John S. Eastwood and the Control of Water in the West* (University of Oklahoma Press).

Jha, A. 2015. "When you wish upon a star: Nuclear fusion and the promise of a brighter tomorrow." *Guardian* (January 25).

Jones, L. 2014. www.mobileburn.com/23466/news/cool-infographic-shows-every-facet-of-the-iphone-6-production.

Kahler, M. 1998. "Rationality in international relations." *International Organization* 52, 919–941.

Kahn, B. 2015. "3 ways the world's power mix is about to change." *Scientific American* (June 25).

Keen, S. 2015. "Nobody understands debt—including Paul Krugman." *Forbes* (February 10).

Khagram, S. 2004. *Dams and Development: Transnational Struggles for Water and Power* (Cornell University Press), chap. 1.

Kirby, A. 2013. "Is global warming cooler than expected?" *Scientific American* (May 24).

Klinenberg, E. 2013. "Adaptation." *New Yorker* (January 7).

Kolbert, E. 2014. *The Sixth Extinction: An Unnatural History* (Henry Holt and Co.).

Kolbert, E. 2010. "The anthropocene debate: Marking humanity's impact." *Yale Environment 360* (May 17).

Kolbert, E. 2009. "XXXL." *New Yorker* (July 20).

Kriegler, E., et al. 2013a. "Success of climate talks vital for 2°C target." Potsdam Institute for Climate Research.

Kriegler, E., et al. 2013b. "What does the 2°C target imply for a global climate agreement in 2020?" *Climate Change Economics* 4, doi: 10.1142/S2010007813400083.

Krugman, P. 2015. "Nobody understands debt." *New York Times* (February 9).

Krulwich, R. 2013. "What happened on Easter Island—a new (even scarier) scenario." NPR online article (December 10).

Kupchan, C. A. 2012. "The world in 2050: When the 5 largest economies are the BRICs and us." *Atlantic* (February 17).

Lagorio, C. 2007. "The most polluted places on Earth." CBS (June 6).

Landes, D. S. 1969. *The Unbound Prometheus: Technological Change and Industrial Development in Western Europe from 1750 to the Present* (Cambridge University Press).

Lean, G. 2015. "Winners and losers in our warming world." *Telegraph* (July 24).

Lean, G. 2014. "Global warming: China and US in crucial talks on cutting carbon dioxide emissions." *Telegraph* (May 2).

Lecompte, C. 2014. "Is nuclear power ever coming back?" *Atlantic* (June 24).

Leddra, M. 2010. *Time Matters: Geology's Legacy to Scientific Thought* (Wiley-Blackwell).

Lelieveld, J. 2015. "The contribution of outdoor air pollution sources to premature mortality on a global scale." *Nature* 525, 367–371.

Levitus, S., et al. 2012. "World ocean heat content and thermosteric sea level change (0–2000 m), 1955–2010." *Geophysical Research Letters* 39, doi: 10.1029/2012GL051106.

Locher, D. A. 2001. *Collective Behavior* (Pearson).

MacLeod, N. 2015. *The Great Extinctions: What Caused Them and How They Shape Life* (Firefly Books).

Magill, B. 2015. "Offshore wind power may finally blow into U.S. waters." *Scientific American* (June 18).

Majerus, M. E. N. 2009. "Industrial melanism in the peppered moth, *Biston betularia*: An excellent teaching example of Darwinian evolution in action." *Evolution, Education and Outreach* 2, 63–74.

Marchal, V., et al. 2011. "OECD environmental outlook to 2050." OECD (November).

Marren, P. 2009. "Professor Michael Majerus: Geneticist who defended Darwin in the battle against creationism." *Independent* (February 13).

Martin, C., M. Chediak, and K. Wells. 2013. "Why the U.S. power grid's days are numbered." *Bloomberg Business* (August 22).

Mason, S. F. 1962. *A History of the Sciences* (Macmillan).

Master, C. 2009. "Media insiders say internet hurts journalism." *Atlantic* (April).

McBride, J. P., et al. 1978. "Radiological impact of airborne effluents of coal and nuclear plants." *Science* 202, 1045–1050.

McCarthy, S. 2015. "Water scarcity poses economic and security threat around the world." *Globe and Mail* (November 30).

McCarthy, S., S. Chase, and B. Jang. 2014. "Canadian government approves Enbridge's controversial Northern Gateway pipeline." *Globe and Mail* (June 17).

McElfresh, M. 2015. "Power grid cyber attacks keep the Pentagon up at night." *Scientific American* (June 8).

McGrath, M. 2015. "Global warming increases 'food shocks' threat." BBC News (August 14).

McGrath, M. 2014. "Climate inaction catastrophic—U.S." BBC News (March 31).

McGuffie, K., and A. Henderson-Sellers. 2013. *A Climate Model Primer* (John Wiley and Sons).

McGugan, I., and R. Younglai. 2015. "Super slump: Why there's no end in sight as resources rout gathers steam." *Globe and Mail* (November 27).

McNeil, D. G. 2010. "Virus ravages cassava plants in Africa." *New York Times* (May 31).

Miller, J. W., and R. Smith. 2014. "The future of coal: Despite gas boom, coal isn't dead." *Wall Street Journal* (January 7).

Miller, M. 2015. "Ray Kurzweil: Solar will power the world in 16 years." BigThink.com (accessed September 25).

Misa, T. J. 1995. *A Nation of Steel: The Making of Modern America, 1865–1925* (Johns Hopkins University Press).

Mogil, H. M. 2007. *Extreme Weather* (Black Dog and Leventhal Publishers).

Mokyr, J. 1998. "The second industrial revolution." Northwestern University course notes, available online.

Monastersky, R. 2015. "Anthropocene: The human age." *Nature* 519 (March).

Mooney, C. 2016. "Scientists say humans have now brought on an entirely new geologic epoch." *Washington Post* (January 7).

Mooney, C. 2015a. "The world is off course to prevent 2 degrees C of warming, says energy agency." *Washington Post* (June 14).

Mooney, C. 2015b. "The world's most famous climate scientist just outlined an alarming scenario for our planet's future." *Washington Post* (July 20).

Morelle, R. 2015. "Anthropocene: New dates proposed for the 'Age of Man.'" BBC News (March 11).

Mourdoukoutas, P. 2015. "The ugly side of globalization." *Forbes* (August 23).

Naude, W., A. Szirmai, and J. Haraguchi. 2015. *Structural Change and Industrial Development in the BRICS* (Oxford University Press).

Newman, C. 2004. "Why are we so fat?" *National Geographic* (August).

Nijhuis, M. 2015. "When did the human epoch begin?" *New Yorker* (March 11).

Nijhuis, M. 2014a. "Can coal ever be clean?" *National Geographic* (April).

Nijhuis, M. 2014b. "Is 2 degrees the right limit for global warming? Some scientists say no." *National Geographic* (October 1).

Nordhaus, T., and M. Shellenberger (eds.). 2011. *Love Your Monsters: Postenvironmentalism and the Anthropocene* (Breakthrough Institute).

Nuccitelli, D. 2015. "Overlooked evidence: Global warming may proceed faster than expected." *Guardian* (April 30).

Nye, D. E. 1992. *Electrifying America: Social Meanings of a New Technology, 1880–1940* (MIT Press).

Olshansky, S. J. 2009. "What will our society look like in 2050?" European Health Forum report, available online.

Orcutt, M. 2014. "China's shale gas bust." *MIT Technology Review* (August 12).

Otto, A., et al. 2013. "Energy budget constraints on climate response." *Nature Geoscience* 6, 415–416.

Owen, J. 2010. "New Earth epoch has begun, scientists say." *National Geographic News* (April 6).

Pakandam, B. 2009. "Why Easter Island collapsed: An answer for an enduring question." London School of Ecomonics working paper 117/09.

Palmer, B. 2012. "Global warming would harm the Earth, but some areas might find it beneficial." *Washington Post* (January 23).

Park, M. 2014. "Top 20 most polluted cities in the world." CNN (May 8), http://www.cnn.com/2014/05/08/world/asia/india-pollution-who/index .html.

Patterson, R. 2015. "2015 could be the year of peak oil." Oilprice.com (June 23).

Patton, M. 2015. "Crude oil prices may be headed lower." *Forbes* (September 23).

Pearlman, J. 2014. "World's climate warming faster than feared, scientists say." *Telegraph* (January 1).

Peter, L. J. 1969. *The Peter Principle* (William Morrow).

Pielke, R. 2011. *The Climate Fix: What Scientists and Politicians Won't Tell You about Global Warming* (Basic Books).

Pittis, D. 2014. "Uber economics could be the new cottage industry." CBC News (November 21).

Plumer, B. 2014. "There have been five mass extinctions in Earth's history. Now we're facing a sixth." *Washington Post* (February 11).

Ponting, C. www.eco-action.org/dt/eisland.html.

Powers, B. 2014. "The popping of the shale gas bubble." *Forbes* (September 3).

Press, S. J., and J. M. Tanur. 2001. *The Subjectivity of Scientists and the Bayesian Approach* (J. Wiley).

Prince, M. O., and J. Shoulak. 2015. "What is the future of nuclear power?" *Wall Street Journal* (June 3).

Prosh, E. C., and A. D. McCracken. 1985. "Postapocalypse stratigraphy: Some considerations and proposals." *Geology* 13, 4–5, doi: 1130/0091 -7613(1985)13<4:PSSCAP>2.0.CO;2.

Raj, A. 2014. "Kurzweil: Solar energy will be unlimited and free in 20 years." *Business Insider* (September 22).

Rathi, A. 2013. "Metals in your smartphone have no substitutes." *Conversation* (December 5).

Ray, R. 2010. "Untapped potential." *Hydro Review* (December 17). See www .renewableenergyworld.com/rea/news/article/2010/12/untapped-potential.

Raymond, R. 1984. *Out of the Fiery Furnace* (Pennsylvania State University Press).

Rehmeyer, J. 2012. "Game theory suggests current climate negotiations won't avert catastrophe." *ScienceNews* (October 29).

Repcheck, J. 2003. *The Man Who Found Time* (Perseus).

Revkin, A. C. 2012. "Exploring humanity's evolving 'global brain.'" *New York Times* (December 3).

Ridgewell, H. 2015. "'Antibiotic apocalypse' a step closer, scientists warn." *VOA* (November 20). VOAnews.com.

Rifkin, J. 2015. "We are glimpsing at the outlines of a new economic system." *European* (February 25).

Rifkin, J. 2013. *The Third Industrial Revolution: How Lateral Power Is Transforming Energy, the Economy and the World* (St. Martin's Griffin).

Roach, J. 2014. "We're kidding ourselves on 2-degree global warming limit: Experts." CNN News (November 28).

Rockwell, T. 2011. "Fear of radiation is killing people and endangering the planet too." *Atomic Insights* (April 20).

Rose, D. 2013. "World's top climate scientists confess: Global warming is just a QUARTER of what we thought—and computers got the effects of greenhouse gases wrong." *Daily Mail* (September 14).

Rosen, J. 2014. "The birds: Why the passenger pigeon became extinct." *New Yorker* (January 6).

Rosenkranz, N. Q. 2015. "Albert Gallatin on government incompetence." *Washington Post* (July 20).

Ross, T. 2015. "Green energy subsidies facing the axe." *Telegraph* (July 19).

Safefood. 2009. Where does our food come from? (July). www.safefood.eu.

Scheer, R., and D. Moss. 2012. "Deforestation and its extreme effect on global warming." *Scientific American* (November 13).

Scheiber, N. 2015. "Growth in the 'gig economy' fuels work force anxieties." *New York Times* (July 12).

Schwägerl, C. 2014. *The Anthropocene: The Human Era and How It Shapes Our Planet* (Synergetic Press).

Schwägerl, C., and A. Bojanowski. 2011. "Do humans deserve their own geological era?" *Spiegel Online* (August 7).

Schwartz, J. 2015. "El Niño may bring heat, and rain for California." *New York Times* (August 13).

Science. 2015. Special issue, "Forest Health" (August 21), 800–836.

Scranton, R. 2015. *Learning to Die in the Anthropocene* (City Lights Publishers).

Sharma, R. 2012. "Broken BRICs." *Foreign Affairs* (November/December).

Shukman, D. 2014. "Geo-engineering: Climate fixes 'could harm billions.'" BBC News (November 26).

Sim, D. 2015. "Climate change winners and losers: Which animal species will thrive?" *International Business Times* (October 20).

Smil, V. 2015. *Power Density: A Key to Understanding Energy Sources and Uses* (MIT Press).

Smil, V. 2005. *Creating the Twentieth Century: Technical Innovations of 1867–1914 and Their Lasting Impact* (Oxford University Press).

Smith, N. 2014. "The dark side of globalization: Why Seattle's 1999 protesters were right." *Atlantic* (January 6).

Sneed, A. 2015. "Moore's law keeps going, defying expectations." *Scientific American* (May 19).

Snowdon, C. 2014. "The fat lie." Institute of Economic Affairs (August). https://iea.org.uk.

Snyder, T. 2015. "Hitler's world may not be so far away." *Guardian* (September 15).

Spence, P. 2015. "Almost every economist agrees: Uber makes us better off." *Telegraph.* http://www.telegraph.co.uk/technology/uber/11902310 /Almost-every-economist-agrees-Uber-makes-us-better-off.html.

Steenhuysen, J. 2013. "Drug pipeline for worst superbugs 'on life support': Report." Reuters (April 18).

Steffen, W., et al. 2011. "The Anthropocene: conceptual and historical perspectives." *Philosophical Transactions of the Royal Society A* 369, 842–867.

Stevenson, C. M., et al. 2015. "Variation in Rapa Nui (Easter Island) land use indicates production and population peaks prior to European contact." *Proceedings of the National Academy of Sciences* 112, 1025–1030.

St. Fleur, N. 2016. "Signs of the 'Human Age.'" *New York Times* (January 11).

Stiglitz, J. E. 2007. *Making Globalization Work* (W. W. Norton).

Stix, G. 2012. "Effective world government will be needed to stave off climate catastrophe." *Scientific American* (March 17).

Stout, M. 2016. "The danger of inadvertent war in the next four years." War on the Rocks (November 16).

Stromberg, J. 2013. "What is the Anthropocene and are we in it?" *Smithsonian Magazine* (January).

Sullivan, C. 2013. "Human population growth creeps back up." *Scientific American* (June 14).

Sullivan, G. 2014. "Earth's ozone layer is recovering." *Washington Post* (September 11).

Sun, L. H., and B. Dennis. 2016. "The superbug that doctors have been dreading just reached the U.S." *Washington Post* (May 26).

Sutter, J. D. 2015. "2 degrees: The most important number you've never heard of." CNN News (May 11).

Taylor, J. 2015. "'Global warming the greatest scam in history' claims founder of Weather Channel." *Daily Express* (June 9).

Teitelbaum, M. S. 2015. "The truth about the migrant crisis." *Foreign Affairs* (September 14).

Teng, F., and S.-Q. Xu. 2012. "Definition of business as usual and its impacts on assessment of mitigation efforts." *Advances in Climate Research* 3, 212–219.

Thompson, D. 2015. "The Uber economy." *Atlantic* (January 23).

Thomson, S. 2014. "Is the recent solar power 'tipping point' a sign of things to come?" *Edmonton Journal* (April 10).

Tierney, J. 2008. "Greens and hunger." *New York Times* (May 19).

Tollefson, J. 2013. "U.S. Electrical Grid on the Edge of Failure." *Scientific American* (August 26).

Tran, A. B. 2014. "Where communities have banned fracking." *Boston Globe* (December 18).

Travis, A. 2012. "Global illicit drug users to rise 25% by 2050, says UN." *Guardian* (June 26).

Ubelacker, S. 2013. "10 million Canadians at risk from exposure to traffic pollution: Researchers." CTV News (October 21).

Uglow, J. 2002. *The Lunar Men* (Farrar, Straus and Giroux).

Usher, A. P. 1988. *A History of Mechanical Inventions* (Dover).

Van der Hoeven, M. 2014. "Solar electricity roadmaps 2014." International Energy Agency Webinar (September 29).

Vaughan, A. 2017. "Solar power growth leaps by 50% worldwide thanks to US and China." *Guardian* (March 7).

Vince, G. 2014. *Adventures in the Anthropocene: A Journey to the Heart of the Planet We Made* (Vintage Digital).

von Kaenel, C. 2015. "Wind power must now contend with extreme weather." *Scientific American* (August 19).

Walsh, B. 2013. "Focusing on industrial pollution." *Time* (November 4).

Walsh, B. 2011. "Today's smart choice: Don't own. Share." *Time* (March 27).

Warner, J. 2013. "No one really understands what's going on in our economy." *Telegraph* (February 11).

Weart, S. 2015. "General circulation models of the atmosphere." In *Discoveries in Modern Science: Exploration, Invention, Technology.* Trefil, J., ed. Vol. 1, 41–46.

Webb, J. 2016. "Famous peppered moth's dark secret revealed." *Guardian* (June 2).

Weightman, G. 2010. *The Industrial Revolutionaries* (Grove Press).

Weldon, D. 2015a. "When good news is bad news." BBC News (September 3).

Weldon, D. 2015b. "Are global wages about to turn?" BBC News (October 9).

Yang, E.-S., et al. 2006. "Attribution of recovery in lower-stratospheric ozone." *Journal of Geophysical Research* (September), doi: 10.1029 /2005JD006371.

Yeo, S. 2014. "India hints at new focus on consumption-based emissions." November 25. http://www.climatechangenews.com/2014/11/24/india -hints-at-new-focus-on-consumption-based-emissions/.

Yep, E. 2014. "Asian refiners get squeezed by U.S. energy boom." *Wall Street Journal* (January 1).

Younglai, R. 2015. "Rise of 'sharing' services Uber, Airbnb points to a precarious labour climate." *Globe and Mail* (October 26).

Zakaria, F. 2016. "Populism on the march—why the West is in trouble." *Foreign Affairs* (October 21).

Zalasiewicz, J. 2015. "The Earth stands on the brink of its sixth mass extinction and the fault is ours." *Guardian* (June 21).

Zalasiewicz, J., et al. 2010. "The new world of the Anthropocene." *Environmental Science and Technology* 44, 2228–2231.

Zimmer, C. 2014. "Century after extinction, passenger pigeons remain iconic—and scientists hope to bring them back." *National Geographic* (August 30).

Zimmer, C. 2013. "Evolution in color: From peppered moths to walking sticks." *National Geographic* (October 9).

Zimmer, D. (ed.). 2004. *4 Way Street: The Crosby, Stills, Nash and Young Reader* (Da Capo).

Zolfagharifard, E. 2014. "'We're f*****': Climate change will be catastrophic for mankind after study reveals methane leaking from the Arctic Ocean, scientist warns." *Daily Mail* (August 8).

Index